Signs, Genres, and Communities in Technical Communication

M. Jimmie Killingsworth

and

Michael K. Gilbertson

Baywood's Technical Communications Series

Series Editor: JAY R. GOULD

Baywood Publishing Company, Inc.

AMITYVILLE, NEW YORK

Library of Congress Catalog Number: 92-7420
ISBN: 0-89503-064-0 (cloth)
ISBN: 0-89503-065-9 (paper)

Library of Congress Cataloging-in-Publication Data

Killingsworth, M. Jimmie.
    Signs, genres, and communities in technical communication / M.
Jimmie Killingsworth and Michael K. Gilbertson.
        p.   cm. - - (Baywood's technical communications series)
    Includes bibliographical references and index.
    ISBN 0-89503-064-0. - - ISBN 0-89503-065-9 (paper)
    1. Technical writing.   I. Gilbertson, Michael K.   II. Title.
III. Series: Baywood's technical communications series (Unnumbered)
    T11.K48   1992
    808'.0666- -dc20   92-7420
                                                            CIP

DEDICATION

To Dr. Lewis E. Jones, Jr., and Dr. Barbara E. Gilbertson

who take the time to communicate
and, in the process, heal bodies and souls

Foreword

This book's audience—presumably on one end or the other of a lecture hall, but possibly including the practitioner from industry—faces a discussion of technical writing that complements the "how to" approach by explaining the "wherefore and why." Beginning with first principles, the authors have searched for a philosophical basis for technical communication (applicable to other forms or genres of communication as well) and have found it in semiotics, the study of signs.

Signs are the gauge bosons of rhetoric. Like those force-carrying particles now being discovered by physicists, signs are only tenuously and ephemerally identifiable as actual things, yet they mediate all interactions among intellects. As mediators themselves—situated between a specialized field of knowledge and a reader or user who may be unfamiliar with the field—technical communicators should appreciate this choice of subject matter.

Killingsworth and Gilbertson analyze and distill the content of semiotics and show how it pertains to technical rhetoric. Hardly the least of their contributions is their exposition and interpretation of Charles Sanders Peirce's general theory of signs, which was originally set forth in fragmentary writings ranging from challenging to impenetrable. The difficulty stems in part from the philosopher's singularly turbid writing style, but to a great extent it is inherent in the philosophical level at which semiotics operates. Semiotics is so basic that explaining it to a writer is like explaining water to a fish. This book opened my eyes and enabled me to see, for the first time, what was all around me and should have been perfectly obvious from the start. Here we have not a set of empirical rules or an aesthetic critique of prose styles, but rather a theoretical foundation for what we do: a foundation that is more basic than Aristotelian rhetoric and the ongoing amplifications thereof, more general than the mechanistic analogies to electrical engineering and information theory, and more readily applicable to technical and commercial writing than are the various fiction- and poetry-oriented schools of criticism.

But Killingsworth and Gilbertson do not rest satisfied with their application of Peirce's particle semiotics; they place this elemental discussion within the generic, social, and historical contexts of technical communication. The analysis of communities and groups that identify their members by special patterns of language use—groups such as social classes or academic specializations—reveals a surprising complexity and a number of ironies. In the early history of technical writing, for example, scientific authors emerging from the middle classes deliberately rejected certain extremes of language, but comparable extremes are now associated with the middle class, which comprises most practitioners and users of technical communication.

These practitioners (many of whom have little formal training in the discipline and may not even think of themselves as technical writers) sometimes find themselves at odds with their reviewers over tiny but revealing items that neatly crystallize this viewpoint. "The language of Artisans, Countrymen, and Merchants," which Britain's Royal Society recommended to technical authors during the early stages of the Enlightenment, might not be embraced by a public that refers to garbagemen as sanitation engineers and to the personnel department as Human Resources. For that matter, can "all amplifications, digressions, and swellings of style" be rejected by a Corporate Communications Specialist or a Senior Associate Information Developer? The language of the aristocracy did not suit the Royal Society's ends, but the language that has since been adopted by the corporate *petit bourgeois* is not necessarily an improvement. Many of today's "merchants," in the larger sense of dress-for-success businesspeople, would be aghast at being lumped together with "artisans" and "countrymen" in language or anything else.

As Paul Fussell and many others have observed, class differences are most clearly defined by their rhetorical signposts. An understanding of this is useful to the writer as a guide to strategies and tactics, which can be based on awareness of the reader's class and its linguistic preferences. Class characteristics also apply to reviewers and are therefore useful as an explanation for what is happening in the review process. (This application must of course be pursued with caution, since the reviewers might well be unaware of, and touchy about, the basis for their judgment. If, say, a marketing manager insists on using wording that is ethically and aesthetically objectionable in order to soft-soap a product's deficiency, waving Fussell's book *Class* in his face is not likely to improve the author's lot.)

In their blending of social theory and semiotics, Killingsworth and Gilbertson provide a comprehensive and provocative theory of technical rhetoric. (They even offer a plausible, Peircean answer to that stubborn and difficult question, "What is technical writing?") The book can be appreciated by the relatively new student, selected by the professor without the shame of reductionism, and applied profitably by the seasoned professional.

Joe Chew
Lawrence Berkeley Laboratory

Acknowledgments

This book arises out of an extended conversation about the theoretical implications of technical communication. The conversation, which began in 1984, when the authors were colleagues in the technical communication program at New Mexico Institute of Mining and Technology, became fairly quickly a project in collaborative research and writing. Many people have joined us in our conversations and have helped us gather information and put it to use in written form.

Above all there is Joe Chew of Lawrence Berkeley Laboratory, who, in addition to authoring the Foreword, has been a faithful reader, critic, and sometimes co-author and editor throughout the project. Jim Corey also had a special role in the development of this study, since he was the department chairman at New Mexico Tech who brought us together in an excellent undergraduate program and urged us to pursue our theoretical work, work that others may have thought too esoteric to be a proper pursuit for technical writing teachers.

Other friends, colleagues, and associates have read portions of our manuscript, have discussed the key concepts with us, or have encouraged us in various ways. They include Valerie Balester, Tommy Barker, Dave Carson, Pat Connors, Don Cunningham, Kathy Ferrara, Jeanette Harris, Barbara Johnstone, Jeff Jones, Ken Ketner, Sherry Little, Mary Sue MacNealy, Larry Mitchell, Bill O'Donnell, Jackie Palmer, Mary Ellen Pitts, John Powers, Carolyn Rude, Scott Sanders, Jeffrey Smitten, Sherry Southard, Beth Tebeaux, and Carole Yee. We also appreciate the great number of students who have read and commented on various versions. Especially helpful have been the students in graduate seminars on technical communication theory at Memphis State University and the graduate seminars on rhetoric and composition at Texas A&M University.

Jay Gould has provided excellent editorial advice and strong encouragement in the latter stages of the project. The book could not have been done in this form without his help, and we will always be grateful for his patience and wisdom—not to mention his endurance! Two anonymous reviewers also provided a great deal of useful advice and commentary on the manuscript. Finally, Stuart Cohen and the fine staff at Baywood have made the production work on the book a pleasure rather than a chore for us.

Parts of the book have been published in earlier versions.

A small portion of Chapter 3 appeared originally in an article by Killingsworth and Scott Sanders, "Compensation and Complementarity in the Design of Technical Documents," in *The Technical Writing Teacher, 17,* pp. 204-221, 1990.

An early version of Chapter 5 appeared in an article by Killingsworth and Gilbertson, "Rhetoric and Relevance in Technical Writing," in *Journal of Technical Writing and Communication, 16,* pp. 287-297, 1986.

An early version of Chapter 6 appeared in an article by Killingsworth and Lynn Waller, "A Grammar of Person in Technical Writing," in *The Technical Writing Teacher, 17,* pp. 26-40, 1990.

Some of Chapter 7 appeared in Killingsworth's "Guest Editorial: Thingishness and Objectivity in Technical Style," in *Journal of Technical Writing and Communication, 17,* pp. 105-113, 1987.

Some of Chapter 9 comes from an article by Killingsworth and Betsy Jones, "Division of Labor or Integrated Teams: A Crux in the Management of Technical Communication?," in *Technical Communication, 36,* pp. 210-221, 1989.

A longer, somewhat more technical version of Chapter 10 was published by Killingsworth as "Realism, Human Action, and Instrumental Discourse," in *Journal of Advanced Composition, 12,* pp. 171-200, 1992.

We thank the editors and publishers of this material for permission to re-use it.

Table of Contents

Lists of Figures
and Tables

Tables   **Page**

CHAPTER 1

Introduction:
A Three-Part Theory of
Technical Communication

THE FAMILIAR AND THE UNFAMILIAR

Over the last few decades, as new technologies have become more readily available and more efficiently adapted to an increasing number of activities and users, the field of technical communication has grown considerably. New writers, teachers, and researchers have appeared to meet the growing demand for people who can facilitate technological actions by communicating specialized information to nonexpert readers in business, government, education, and the home.

An Autobiographical Note

One consequence of this growth is that the circle of technical communicators has generously expanded to include more and more academic people. Among this group were literally hundreds of English professors, including the authors of this book.

With the expansion of English courses and programs in technical writing, we found ourselves crossing over from literary and rhetorical studies to teach the new courses and to work as consultants in the busily growing field. The change proved to be enlightening.

In putting us on relatively unfamiliar ground, the work required us to take stock of our resources—what we already knew from literary and rhetorical scholarship—and to reconcile that knowledge to the new knowledge we were developing daily in our study of how technical writers work and learn. Faced with an unfamiliar situation, then, we discovered how previous experience prepared us for the work (or failed to prepare us) and how, in turn, our new work tempered our previous knowledge with fresh insights and understanding.

The Value of the Unfamiliar

Reconciling the unfamiliar to the familiar is, of course, a cognitive problem that technical writers frequently confront. Writing a users manual for a computer program, proposing a technical project to a funding agency, writing a feasibility study for a new kind of technology—all such activities require writers to carry their audience away from a comfortable grounding in familiar experiences and lead them gently but surely onto unfamiliar terrain.

The relation of the familiar to the unfamiliar is also a concern for the writers of fiction and poetry, but they come at the problem from the other direction. Literature, especially romantic and modern literature, tends to throw readers into unfamiliar situations and to stimulate unfamiliar emotions and thoughts, gradually bringing the reader to recognize the core of familiarity within the unfamiliar [1]. If, for example, readers begin Franz Kafka's story, *The Metamorphosis*, dismayed over the thought that, in his sleep, the protagonist has turned into a giant cockroach, they are likely to finish the story moved by a strange sympathy, having recognized the general human condition of sufferers like Gregor Samsa, who proceed buglike through the world, always worrying that some figure of admiration or authority—father, boss, or lover—will step on them and destroy their lives.

Recognizing that literature creates important effects by this process of "defamiliarization," theorists in the study of language and literature have borrowed the technique. They have found that making initial demands on readers may force these readers out of an habitual way of thinking, giving them a fresh perspective and allowing them to return to the everyday practice of reading, writing, thinking, and living, with eyes refreshed by insight.

Discourse Theory and Defamiliarization

That, as we see it, is the aim of discourse theory: In making familiar language practices temporarily unfamiliar, it allows us to get some distance on our practices and thereby to question those practices, put them to the test, and ultimately to improve them. Theory creates distance on habit, allows us to see it as others see it—which is the only way to reform habit.

In this book, which applies discourse theory to a study of technical communication, we want to re-create for people already familiar with technical communication something of the sensation of unfamiliarity we experienced upon entering the field. We will put these readers on the unfamiliar ground of discourse theory and guide them on their gradual journey back to the familiar advice of textbooks and conference workshops. We want them to stand back from the busy world of practice and look at technical communication as it relates to "the big picture" implied in the question, what does it mean to communicate?

Beginning with a general theory of signs—the basic elements of communication, so basic that they seem unfamiliar—we proceed to the more familiar ground of genre theory, dealing with generalizations about reports, manuals, and proposals, the types of technical communication. Genre studies have informed instruction in technical communication at least since the 1920s [2], but we want to look at the genres in two relatively new ways, coming at them first from a background in the study of signs and then proceeding from them to a study of discourse communities. Following Carolyn Miller [3], we interpret genres as forms of social action that are illuminated by relevant—though perhaps unfamiliar—concepts from a social theory of communication. In challenging technical communicators to search for an understanding of their work in terms of discourse and society, we ask another big question, what is the *place* of technical communication in a technological culture as a whole?

If technical communicators find themselves "at sea" in this large realm of generalization, we hope gradually to lead them back to familiar ground, to create an AHA! experience that helps them to see their work and themselves in a new light. Having thus returned, we hope they will be enriched by the experience of travelling beyond the daily grind of workplace custom and previous experience.

Of course, some readers will come to us from the other side, from the academic aisle of technical communication. For these readers, we hope to demonstrate the great value of technical communication as a ground for applying and questioning our most cherished theories of discourse. If theorists work exclusively with literary and philosophical texts, they can only produce a partial theory of written communication. To move toward a comprehensive theory requires that we examine texts and practices from outside the walls of the library and classroom. If our teaching takes us into the realm of practical discourse anyway, we ought to carry what we have learned from the study of literature with us, applying it where possible and questioning it when it doesn't seem to fit our needs.

Still other readers, especially students in the field, will come to this book untrained in both discourse theory and technical communication. To them, we hold out this hope: The struggles through the unfamiliar terrain of the early chapters will be rewarded in the late chapters when we return technical communication to its social context, the academic and technological communities in which modern people live and work.

This act of returning is perhaps the most important thing that theory can do. It situates language use within larger historical and social contexts, describing what we do in relation to where and when we do it, and thereby providing a descriptive map that relates discourse to action.

Though we are not so vain as to think that the field of technical communication has "matured" in recent years simply because it has made room for more academic practitioners, we do hope that the field has benefitted from the infusion of English teachers and students. In addition to initiating formal, empirical research in the field, literary scholars and rhetoricians have other, more traditional contributions to make. These include scholarship—the reading and study of texts—and discourse theory, one of the more recent permutations of scholarship in literature and language. In this area, we hope to make our mark, giving back to technical communication some of what it has given to us.

Why Theorize?

Our purpose, then, is to produce a descriptive theory, a set of generalizations that map the territory of technical communication. If successful, this theory should serve teachers as a framework for organizing courses on technical communication. It should serve technical writers as a means of relating one project to the next. It should serve students as a way of relating unfamiliar material (what they are learning about technical communication) to familiar material (what they already know about other kinds of writing or about the world in general). Finally, our theory may serve empirical researchers as a source of hypotheses to be tested in quantitative and qualitative research in the field.

What is the exact nature of our generalizations? Scholars in English studies these days, the people who teach and study rhetoric, composition, literary criticism, and linguistics, use the word *theory* in a way different from scientists [4]. In science, theory is a body of testable propositions that may, in principle, be satisfactorily verified or falsified under specified circumstances.

In the humanities, however, such certainty, though it may ideally exist, is not practically attainable, mainly because of the radical individuality and unpredictability of the human mind. The theory of gravity, which states the relation of physical objects to one another, will produce successful predictions far more often than a theory of authorship, for example, which attempts to state how the relationship of one human being to another is mediated by the written word.

Why theorize, then, if our theoretical statements cannot be adequately tested? The main reason is that we can't help it. Human beings always generalize about individual experiences. If we generalize inadequately, we fall into stereotypes or into ineffective repetitions of experiences. If we generalize effectively, we gain a measure of control over our actions and improve our productivity.

For example, if a writer generalizes that writing is an individual struggle of the human mind to express itself in a linguistic medium that is neither subtle nor flexible enough to render the idea in a pure form, that writer is possessed (consciously or unconsciously) of a Platonic or a romantic theory of discourse. This theory secretly informs the thinking of anyone who says that it is possible for a person to know what's in the mind without knowing how to say it.

There are alternatives to this way of looking at things. But a person committed to a particular concept of writing and equally committed to the idea that theory is irrelevant to the practice of writing will not be able to see the alternative—will not be able to see, for example, that a Platonic and individualistic model of authorship could be replaced with a social or collaborative model [5].

The social model says that meaning does not necessarily exist in the mind if it cannot be expressed; instead meaning emerges in dialog and discourse, in the interchange of information among two or more people. The shift in theories could have practical effects upon our hypothetical writer, who, having shifted theoretical gears, is now more likely to consult colleagues about an informational problem instead of sitting for hours struggling with the problem of expression.

In other words, all writers have a theory—whether they know it or not—because all writers tend to generalize from their experience, and their generalizations inevitably influence the way they write. By becoming conscious of the power and the limits of their personal theories, they can open themselves to change and thereby to growth and improvement. By staying steadfastly unconscious—by saying, for example, that theory is irrelevant or nonexistent—they limit themselves to the familiar, the habitual.

So long as the habitual way works, everything is all right. But the world changes, and the adaptable writer survives. One tool of the adaptable writer is theoretical understanding. Theory can make sense of all the little rules and prohibitions that writers customarily draw upon. It can enable communicators at once to comprehend and to question common sense, trial-and-error practice, and a loose sense of convention.

Theory does not necessarily reveal particular points of practice that have not been guessed before, but rather shows how one point can be connected to another in a principled way. It thereby gives a rationale for practice—a rationale that the writer can use, for example, to explain and justify actions to colleagues, something more to say than "I like it that way" or "That's the way it's done."

SIGNS, GENRES, AND COMMUNITIES: INTERRELATED ELEMENTS OF THEORY

We are certainly not alone in turning to discourse theory as a fruitful source of insights on technical communication. A number of writers have successfully applied rhetorical theory [6, 7], and, more recently, excellent studies have appeared in speech-act theory [8-14] and semiotics [15, 16].

All of these theorists recognize that not just any theory will do in technical communication. Classical rhetoric and speech-act theory are more useful than some other theories—deconstructionist poetics, for example—because they show the way between writing and action. As Timothy Weiss has noted, "teachers and theorists of technical and professional communication need a theory . . . that refers to the world outside of discourse" [16, p. 38], that plots a course between writing, reading, and action. Like Weiss, we find that a blend of semiotics and social theory fills this need nicely.

Signs

In the chapters that follow this introduction, we take our initial cue from the American philosopher and scientist, Charles Sanders Peirce (1839-1914), one of the first thinkers to offer a general theory of signs. Peirce's theory is particularly appropriate to technical communication because it is based on philosophical pragmatism, a movement founded by Peirce.

Pragmatism holds that meaning arises from the interplay of thoughts, signs, and action. No ideas, according to this viewpoint, are pure; all are the product of communication and must take some form in images or language before they can really be said to exist at all. Ideas, then, are texts. We don't just think them; we write and read them in our minds.

According to Peirce, we best understand the meaning of a text (whether mental or actual) when we can project the effects that this text will have upon our actions. We read in order to act. We write in order to help someone else to act. The idea of purpose is best understood if it is connected to the action that will be produced by writing. The purpose of a computer manual, for example, is not adequately accounted for by the commonly used "aims" or "modes" of discourse—to "inform" or "persuade" the reader, or to "describe" reality, or to "express" the author's idea, or to "tell a story." Rather the purpose is to ensure the proper use of the computer, to help the reader act in the right way.

Peirce claimed that every such exchange between writing and action involved two key processes—representation and interpretation, the production and consumption of signs. In his way of thinking, there is no transparent communication, no understanding unmediated by signs. All writers and all readers must both interpret and represent information before they can act in a rational way.

The two chapters in Part I analyze these concepts in detail and build a rudimentary theory of document design upon the foundation provided by Peirce's general theory of signs.

Genres

In Part II we explore how signs—the elements of representation and communication—combine into *codes*, elaborated systems of signs that writers and readers use to cooperate in technical actions. One way of categorizing codes, drawn from rhetorical theory, is to classify types of writing, or *genres*.

In our treatment of the major genres of technical communication—the report, the manual, and the proposal—we discover a familiar island of theory, which can be approached from the elemental perspective of semiotics or from the broader perspective of social theory. Genres are conventional ways that writers and readers represent and interpret actions so that they come together to create mutually beneficial projects.

In our scheme, then, genres provide a stopover, a point of connection between the study of signs and the study of communities. The middle chapters dwell upon this familiar ground before heading off again into the deeper water of the social theory of discourse.

Communities

Like signs and genres, communities are communication *media* in the sense that to communicate, writers and readers must transmit information *through* them. Our argument throughout the book is that any medium of communication will affect the manner and the meaning of the message delivered through it.

Thus, a particular kind of sign—a line drawing, for example—will affect the way the reader responds to the message it represents. The message would have a considerably different effect if it were given in words. Since, in pragmatism, the meaning is tied directly to an effect upon action, and since different kinds of signs are likely to change the way a reader responds to a text by acting or failing to act, the sign can be said to have altered the meaning of the message. In this way, the study of signs in pragmatic semiotics goes beyond traditional semantics, which tends to divide more radically media effects and message effects. Semiotics denies this strong separation, largely because of the pragmatic insistence that meaning cannot be separated from action.

Just as a change in the sign type can affect the meaning or the effect of the communication, so can a change in document type or genre. A proposal puts information to use in a way different from a manual. Likewise, a change in the community of discourse creates a major change in the effect a document will have. The same technical scientific paper that has no meaning for an untrained lay reader could be meaningful enough to change the entire research program of a specialized field of science.

Thus, in Part III, we look to social theory for enlightenment about how writers and readers interact in this third major media or context for communication—the discourse community. Figure 1 gives a simple diagram to show how the three parts of our study relate the three types of communication media to three bodies of theory.

Interrelations

The three parts of the theory—signs, genres, and communities—are thus deeply interrelated.

They are related theoretically because they are all kinds of media, even though to study them means to draw upon three distinct bodies of theory: Semiotics covers signs, rhetoric covers genres, and social theory covers communities of discourse.

They are related practically. Any time a writer goes to work, and any time a reader reads, all three media present themselves and must be negotiated before action is possible.

Our thesis, then, is that technical action is mediated by at least three major kinds of influences—signs, genres, and communities. Technical communication is especially concerned with how writers manipulate these media so that readers can find their way from the world of the text to the world of action.

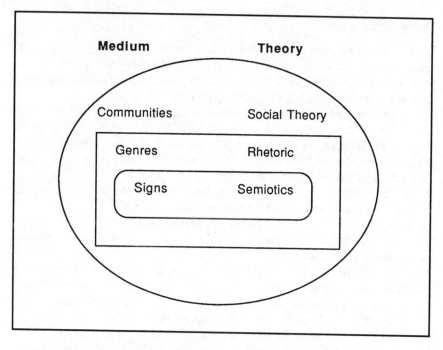

Figure 1. The interrelation of signs, genres, and communities
of discourse.

REFERENCES

1. R. H. Stacy, *Defamiliarization in Language and Literature,* Syracuse University Press, Syracuse, New York, 1977.
2. R. J. Connors, The Rise of Technical Writing Instruction in America, *Journal of Technical Writing and Communication, 12,* pp. 333-337, 1982.
3. C. R. Miller, Genre as Social Action, *Quarterly Journal of Speech, 70,* pp. 151-167, 1984.
4. C. R. Campbell, What Can Discourse Theory Offer the Professional Who Writes?, *IEEE Transactions in Professional Communication, 33,* pp. 156-161, 1990.
5. K. B. LeFevre, *Invention as a Social Process,* Southern Illinois University Press, Carbondale, 1987.
6. R. E. Masse and M. D. Benz, Technical Communication and Rhetoric, *Technical and Business Communication: Bibliographic Essays for Teachers and Corporate Trainers,* C. H. Sides (ed.), National Council of Teachers of English, Urbana, Illinois, pp. 5-38, 1989.
7. S. Dragga and G. Gong, *Editing: The Design of Rhetoric,* Baywood Publishing Company, Amityville, New York, 1989.
8. M. P. Haselkorn, Linguistic Boundaries of Technical Writing, *Technical Writing Teacher, 11,* pp. 26-30, 1983.

9. M. P. Haselkorn, A Pragmatic Approach to Technical Writing, *Technical Writing Teacher, 11*, pp. 122-134, 1984.
10. K. Riley, Conversational Implicature and Unstated Meaning in Professional Communication, *Technical Writing Teacher, 15*, pp. 94-108, 1988.
11. K. Riley, Pragmatics and Technical Communication: Some Further Implications, *Technical Writing Teacher, 13*, pp. 160-170, 1986.
12. K. Riley, Speech Act Theory and Degrees of Directness in Technical Writing, *Technical Writing Teacher, 15*, pp. 1-29, 1988.
13. R. W. Dasenbrock, J. L. Austin and the Articulation of a New Rhetoric, *College Composition and Communication, 38*, pp. 291-305, 1987.
14. D. N. Dobrin, *Writing and Technique*, National Council of Teachers of English, Urbana, Illinois, 1989.
15. B. F. Barton and M. S. Barton, Simplicity in Visual Representation: A Semiotic Approach, *Journal of Business and Technical Communication, 1*, pp. 5-26, 1987.
16. T. Weiss, Bruffee, The Bakhtin Circle, and the Concept of Collaboration, *Collaborative Writing in Industry*, M. M. Lay and W. M. Karis (eds.), Baywood Publishing Company, Amityville, New York, pp. 31-48, 1991.

Part I
SIGNS

CHAPTER 2

A General Theory of Signs

SIGNS AND ACTION IN
TECHNICAL COMMUNICATION

In the technical lingo of semiotics, a sign is a *representation*; it stands for something else. It is not a new concept. In ancient philosophy, a sign was defined as *a presence that stands for an absence*. Thus, the word *tree* can stand for the actual but absent physical entity with branches and bark. Or we could say that the word *tree* stands for the idea of the tree in the mind of a sign user (speaker, writer, listener, or reader).

Words are not the only signs, though. A drawing is also a sign, since it could stand in place of the thing or the idea, as could finger motions in sign language. In addition, physical evidence of a tree—an acorn or a leaf—would be a sign, an indication that a tree is present nearby, perhaps over the next hill. In this way, a cough is a sign of a cold, and smoke is a sign of fire.

Just how useful the concept of the sign could be to communication theory is suggested by the philosopher Max Fisch. Fisch says that a general theory of signs should "give us a map so complete and so detailed as to place any field of highly specialized research in relation to any other, tell us quickly how to get from one such field to another, and distinguish fields not yet explored from those long cultivated" [1, p. 360].

Technical communication is one such field. It has developed in the interstices between highly developed scientific and technological disciplines. As these fields grow ever more specialized and their knowledge becomes ever more formidable from the perspective of other fields, they begin to need the kind of writing that preserves their connection to the "outside world" for purposes of funding, recruiting new practitioners, selling their products, teaching, advising, and engaging in other generally communal activities. A general theory of signs that sketches paths among expert fields and between any specialized field and the common ground of public discourse offers a logical starting place for a theory of technical communication.

Signs and Human Action

The concept of the sign is thus broad enough not only to apply to various fields of communication practice but also to include all the elements of a technical text: words, pictures, graphics, headings, even white space.

In addition, the concept of the sign allows us to discuss the relationship of thought, text, and action in a special way: *Ideas, texts, and action may have a similar, if not the same, structure.* They may all be signs. As the Russian semioticist Volosinov writes: "any item of nature, technology, or consumption can become a sign, acquiring in the process a meaning that goes beyond its given particularity" [2, p. 10].

People attribute meaning not only to words, drawings, and sign language, but also to things and actions. Americans, for example, have chosen to give a profound meaning to a piece of colored cloth, decorated with stars and stripes. (Even to describe *the flag* in such mundane words may appear as a sign of disrespect for an object protected by the law of the land!) In a different, but related, kind of sign-reading, archaeologists have inscribed meaning upon bone chips found in remote sites where, in prehistoric times, early humans were likely to have lived. The flag and the bone chips are nothing more than themselves until the sign users give them meaning—*sign*ificance. Through the act of interpretation, patriots and scientists connect these things with actions of the past—the heroic deeds of war ("bombs bursting in air") or the activities of early human cultures—giving the things a mythic or scientific significance that, in the words of Volosinov, goes beyond their given particularity.

Moreover, meaning is not only *attributed* to things and actions through interpretation. Meaning is also *conveyed* through things and actions. A person who fails to greet an office mate early on Monday morning may be signifying a bad mood. Actions speak for mental states just as surely, though not perhaps as precisely, as any sentence could. Thus, actions as well as words can be constructed and then "read" as signs.

Clearly enough, the texts of technical communication are systems composed of signs; they are designed to represent a reality and to be interpreted by knowing readers. But, if we take a semiotic view of the relation of texts and technology, we can say that *technical actions are themselves also signs and are a part of the same sign systems that include the texts.*

Like any other form of human action, technical activities may be used as interpretations or representations. In either case, they become signs. These actions may have been made possible by a text, say a technical manual. The manual tells people how to do something in a particular way; it maps an activity. Just as surely as the prose of the manual represents ideal actions, the actions performed by the user of the manual are therefore representations of the text. If the actions are done as the manual suggests they should be done, then the actions may be considered an accurate representation of the manual's text. If the manual says, "Now enter the data in field A," and the user, at the proper moment, succeeds in entering the data, we infer a logical connection between the command and the act; that connection is what we mean by representation. If we observe the user reading the manual and then entering the data, we can infer that the manual tells the user to behave in that way.

Thus the action, from our point of view, is a sign, *a presence that stands for an absence.* We read the action as a sign much in the same way that we draw connections between a photograph and its subject or a word and its referent.

The very idea of "usability testing" depends upon the belief that actions represent a text. A usable manual is one that leads the user from reading to acting. The person giving the test infers that acting and reading are connected in the successful operation of a sign system consisting of the manual and the product.

Writing and action are constantly changing places during the cycle of a typical technological project. Usability testers can write a report based on their observations, for example. The report retranslates the actions into words and pictures once again. The manual text that had prompted the action of the user now in turn gives rise to the report text.

There is thus a semiotic interplay between texts and technical actions. Texts and technology are interchangeable signs, representations and interpretations of one another.

The Sign as an Active Presence

Signs, then, are words, images, things, or actions that stand for other words, images, things, or actions. The flexibility of the sign trends toward tautology; it becomes a concept that can only be defined in terms of itself.

In this sense, the flexibility and breadth of the concept threaten its usefulness. In becoming everything, the sign tends to become nothing. To take an analogy from theology, we can compare the breadth of the sign concept to the concept of the immanent God. If God is in everything, why not talk about things and leave God out of it? If, however, God's presence is active—if God affects things in a special way—then we cannot talk about things without accounting for His effect upon them.

As in theology, so in semiotics: What is needed is a good account of how signs affect the way the world works. We need a theory of the signs as an active presence.

C. S. Peirce's "Pure Rhetoric"

Establishing such a theory was one of the many projects of the turn-of-the-century American philosopher Charles Sanders Peirce (pronounced like "purse"). Now recognized as one of the co-founders of modern semiotics (the other being the contemporaneous Swiss linguist Ferdinand Saussure, who independently developed a different system of sign analysis), Peirce has only recently been studied for his contributions to rhetorical theory (most notably in the work of Lyne [3]).

Peirce had a special interest in rhetoric. As a practicing scientist and logician, as the founder of the school of philosophy known as pragmatism (a word he coined), and as a prolific writer and lecturer on topics from many fields, Peirce may well have been the first theorist of technical communication.

He is thus a figure with whom all technical communicators should be familiar. Reading his work, however, takes some special training and patience. Very often it is highly technical and filled with a difficult jargon that Peirce developed to talk about his general theory of signs. (The literary theorist Robert Scholes has dryly observed that Peirce was addicted to opium and terminology [4].)

We will try to distill some of Peirce's most important thoughts about signs to make them more generally accessible. In so doing, we hope to show that the theory of communication that may be gleaned from Peirce's published and unpublished writings sheds new light on practical rhetoric, both in technical communication and in other fields.

Peirce's understanding of how signs function is particularly appropriate for technical communication theory because his system of thinking is deeply concerned with the way actions are controlled and reproduced in a community of scientific investigators and technical workers. The task of technical rhetoric, he said, is "to ascertain the laws by which in every scientific intelligence one sign gives birth to another, and especially one thought brings forth another" [5, vol. 2, p. 229]. This interplay of thoughts and signs frequently culminates, according to the pragmatist Peirce, in the actions of a technological community.

We begin with an overview of Peirce's concept of the sign and proceed directly to a Peircean analysis of a typical rhetorical situation in technical communication. By analyzing this situation, we hope to make clear that the crucial point in Peirce's theory of signs is the description of the *process of semiosis*.

Peirce portrayed the action of the sign in the world as a continuous interplay of *representation and interpretation,* interchangeable conversions of mental action to verbal action and verbal action to physical action. Interpretation allows such conversations to occur in any direction or in multiple directions. Physical action may be converted to verbal action (as in a report), verbal action may be converted to physical action (as in a manual), and thinking about potential physical actions may be given a verbal form (as in a proposal).

By admitting that interpretation must always accompany representation— a point neglected or denied in positivist rhetorics of technical communication— we can develop a structural model of communication with richness and flexibility. An examination of this model in graphic form allows us to provide support for many of the cherished beliefs of technical communicators and to place many others in question.

THE ELEMENTAL SIGN

It is difficult to summarize in a sentence or two Peirce's complete understanding of the sign, the basic building block in this model of how we think, communicate, and conduct life in general. In a letter to the British semanticist Victoria Welby, Peirce tried to provide a short definition—with only moderate success: "I define a Sign as anything which is so determined by something else, called its Object, and so determines an effect upon a person, which effect I call its Interpretant, that the latter is thereby mediately determined by the former" [6, pp. 80-81]. This effort to compress the subtlety of the full concept into the space of a single sentence played havoc with syntax; Peirce's intellect was uncompromising in this way.

Let us offer a simplification of the sentence. Then we'll work to restore the subtlety. We can say that *a sign links an object to an interpretant, thereby giving new meaning to the object and altering the interpretant by providing new information to be interpreted.* The word *tree,* for example, is a *sign* representing the green leafy *object* outside the window to the mind of a person who uses (speaks, writes, hears, or reads) the word. The word at least partly determines what the language user will think about when confronted with the word. In the mind, there arises an image of the object or a thought relating to the object. This *effect* in the user's mind is called the *interpretant.*

But, in giving this definition, we begin to sense Peirce's difficulty. For we have not really defined *sign*; we have defined *semiosis,* the process of sign function: *semiosis* is the process by which a sign links an object to an interpretant. If we try to give a classic Aristotelian definition by placing the word *sign* in a class of objects (*genus*) and then distinguishing it from other members of the class (*differentia*), we mire our definition in tautology; we use a word in the definition that is part of what we want to define, thus: a sign is a thought, object, or action, that links another thought, object, or action to a third thought, object, or action so that the first gives new meaning to the second by exposing it to a third that is in turn also altered in some way by the new information provided.

The sign may be a word, an image, a thing, but so may its object and even its interpretant. The relationship is understood to be inviolably *triadic*; that is, one of the three parts of the sign can only be defined and distinguished as it relates to another. No wonder, then, that the definition is tautological.

To compound the tautology, in some of his writings, Peirce uses the word *sign* to describe one part of the triadic relation—the part responsible for representing the object to the interpretant—but in other places he uses the word *representamen* to indicate this function. The sign itself then includes the entire relation of three parts: the object, the representamen, and the interpretant.

To have a word to stand for the full relationship of the three parts, we will use the word *sign* in this way, to refer to the whole triadic relationship of the representamen, the object, and the interpretant. We will keep the word *representamen* to signify only the vehicle or *medium* of representation. Thus, in our simple example, the word *tree* is the representamen that stands in place of the natural object (the thing with leaves and bark) and thereby produces the interpretant, or effect, in the mind of the language user. The interpretant could be a feeling, an image, or a thought of some other kind. The *sign* that we have identified is the entire, inviolably triadic relationship, *Tree*-represents-a-tree-to-a-language-user, in which construction the word *tree* is the representamen, the concept (or thing) is the object, and the language user is the interpretant.[1]

Retaining sign to designate the full relationship helps to emphasize that the three parts of the sign must always work together. They have no real existence without one another. Moreover, Peirce would argue that "this tri-relative influence" is "not in anyway resolvable into actions between pairs" [5, vol. 5, p. 484]. None of the three parts can exist as a monad, an item in isolation, or as a dyad, a relation of two items. Nor can the relationship be thought of as a loose connection of three dyads (relating object to representamen, representamen to interpretant, and interpretant to object).

This suggests that for "tree" to signify the leafy green object in the yard there must be a language user willing to make the connection and able to do so because of a mental "dictionary" of interpretants or possible meanings.

A more radical suggestion—which, as we shall see, Peirce was fully willing to make—is that for the language user to comprehend and understand the object, there must be a word or some other form of representamen. Understanding— indeed the very act of thinking—presupposes signs. Moreover, as we shall show, rational action depends upon the effective functioning of signs. Thoughts, words, and deeds are signs that give rise to other signs of a similar nature.[2]

Figure 2 is a diagram of the sign's irreducible triadic relation. In the diagram the

- *object* is that which is represented in the sign (a leafy green plant in the yard),
- *representamen* is the medium of representation and interpretation (the word *tree*),
- *interpretant* is the effect upon an interpreter, the agent that gives meaning to object and that is in turn "determined" by the object (the effect or meaning produced in the mind of a language user who in recognizing the tree is compelled to name it).

The words *representation* and *interpretation* are crucial. According to Peirce, either the representamen (with its representative function) or the interpretant (with its interpretive function) or the object (with its ability to determine the interpretant, to bring forth the effect of semiosis in the mind of a language user) may be viewed as the active agent in the relation. The descriptions of the elements of the relation, like the definition we offered earlier, seem somewhat tautological; the function of each element can only be described in terms of its partners in the relation. This indicates again the essential *irreducibility* of the triadic sign; it cannot be thought of as a set of separate monads or coordinated dyads.

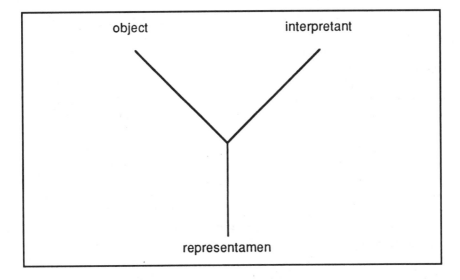

Figure 2. The elemental sign.

The grammar of the diagram, in the formulation Peirce preferred, is this: *the representamen represents the object to the interpretant* ("tree" represents a tree to a language user). But we could also say that *the interpretant interprets the object by means of the representamen* (the language user interprets a perception of a tree as "tree"), or *the object determines the interpretant by means of the representamen* (the tree in the yard stimulates the language user to think or speak the word "tree"). Again, Peirce's formulation of the sign as an irreducible triad suggests that all of these actions are present at once, that every term so qualifies the others that any existence they may have outside of this sign relation would be profoundly different from the one they enjoy within the relation.

COMPLEX SIGNS

Signs, then, are composites of thoughts, texts, and actions that give rise to other such composites. Sign function is characterized by a representamen that connects or identifies an object with an interpretant. The representamen, object, and interpretant are inextricably bound to one another; to alter or simplify their relationship in any way is to create something profoundly different.

Somewhat ironically, the difficulty of explaining and understanding the simple sign is resolved somewhat once we begin to look at complex signs and to recognize how signs function in relation to one another. To think of the sign in isolation may even be misleading. It is most certainly unrealistic. As the linguist

Kenneth Pike notes, "Only if a theory is simpler than that reality which it is in part reflecting is it useful" [7, p. 6].

When we begin to diagram complex signs, then, we move closer to a recognizable reality. The situation is rather like that of atomic physics. A single atom—or worse yet, a single subatomic particle—is a foreign and difficult object to comprehend; but, as we begin to combine atoms into molecules and molecules into chemical substances, we move ever closer to familiar objects that we perceive as existing all around us.[3]

To represent even the simplest version of communication between an utterer and a receiver of a message, for example, we need a diagram not of a single sign (as in Figure 2) but of at least a double sign, like the one in Figure 3. (To make the relationship easier to comprehend, we have identified the interpretant with an interpreter, instead of remaining absolutely faithful to Peirce's understanding of the interpretant as the *effect upon* an interpreter.) The grammar of this relation is this: An utterer (i) interprets the thought (o) represented in the signs of his or her mind (r, which may be words or pictures, but are, according to Peirce, most certainly signs [5, vol. 5, p. 251]). The utterer then represents his or her thoughts at the moment of communication (o'), to a receiver (i') in spoken language (r').

To make it concrete, think of the relationship this way. An utterer (i) is thinking of a tree (r), a tree seen the day before (o). The utterer represents the thought of that tree (o') to a receiver (i') by using the word "tree" (r').

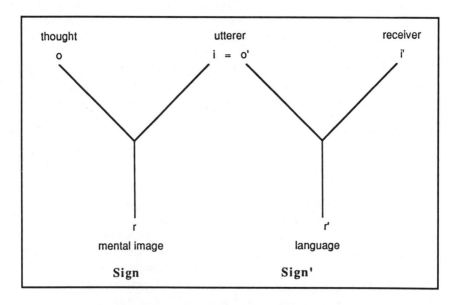

Figure 3. The semiosis of the communication act
(o = object, r = representamen, i = interpretant).

The "=" sign thus indicates a process of transfer, in which the utterer turns from inward thoughts to outward expression. Utterers thus identify themselves with (make themselves equal to) the thoughts they are trying to put forth. They "express themselves."

Let's fill out the example with a couple of characters:

A father sees a tree, recognizes it, and identifies it as a maple for the benefit of his daughter, who is now the wiser for knowing that this particular tree is a maple.

Figure 4 is an attempt to diagram this situation in the simplest way possible.

It may have been a rudimentary version of sign' that stimulated Peirce's first version of the semiotic triad. In an illuminating article, the philosopher Joseph Ransdell notes that Peirce "actually derives the basic object-sign-interpretant relation from the idea of an utterer-utterance-interpreter relation. . . . The most puzzling part of this is no doubt the derivation of the concept of the 'object' from the concept of the 'utterer.' [But] . . . if we strip the concept of the utterer down to

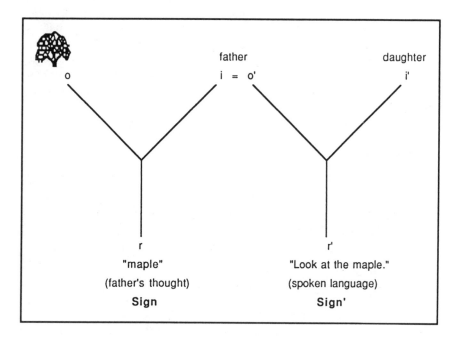

Figure 4. Semiotic diagram of a simple communication situation.

its semiotic essentials it turns out to be simply the general idea of that which is requisite as the basis for the correction of misinterpretation, and that is really what the semiotic concept of 'object' comes down to as well" [8, pp. 172-173].

Thus, in our simple example in Figure 4, o is the tree itself, the *evidence* which could, in the presence of another interpreter, prove the father's identification of it as a maple to be correct. Likewise, the father himself is the object in the communication with the daughter. His existence is the basis for the daughter's correction of her understanding of his statement. If, the next day, she misremembers and refers to the tree as a mulberry, the father's presence guarantees that she could be corrected. Or, if she tells her friend that the tree is a maple, and the friend doubts that she is right, she can call Dad out to verify her information.

The concept of the object as conceived by Peirce and interpreted by Ransdell is ingeniously flexible and far-reaching in its explanatory power. The object subsumes, for example, the rhetorical concepts of both *evidence* and *authority* and shows that they share a similar function. We know how this rhetoric works in scientific and technical communication. A researcher can make a point either by referring to an authority on the subject ("the literature") or by presenting the evidence of original research. If the researcher can do both, all the better. Peirce's semiotic thus accounts for the characteristic processes of verification and falsification in scientific thought and communication.

As we shall see, the ingenuity of Peirce's move lies in his *personification* of the object in the way suggested by Ransdell and in his *depersonalizing* of the interpreter in his concept of the interpretant. There's still a person here, but the interpretant is the *effect* that the representamen has on that person. Peirce was able to generalize the communication relation to show that thought itself is dialogic, like a conversation, and that it is therefore semiotic. "The whole function of the mind," Peirce writes, "is to make a sign interpret itself in another sign and ultimately perhaps in an action . . ." [9, p. 66].

To see how, in the process of communication, "one sign gives birth to another" and how such a proliferation may involve actions as well as thoughts and words, consider the following situation:

> A project engineer needs to get some work done by his technicians. He develops a set of performance objectives and writes an instructional manual for the task. The technicians read the manual and are able thereby to perform the action. Another engineer observes the work and writes a report about it.

Our graph now becomes a chain of triadic relations, as shown in the diagram of Figure 5. The signs of equivalence (i = o) now indicates "valences," symbolic transformations or transactions, points where the effect of the previous object (its

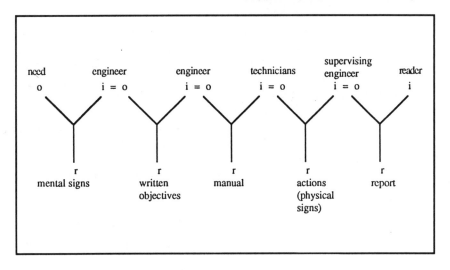

Figure 5. Semiosis in an engineering project.[4]

interpretant) becomes a new object to be represented and interpreted. That inter-pretation produces an effect (interpretant) upon a person, which in turn invites interpretation, and so on—*in a potentially endless chain.*

We can read the diagram this way: The need (to get something done), as represented by mental signs, produced an effect on the engineer. That effect (the engineer's interpretation), as represented by the written objectives, produces a new effect, or a revised interpretation, in the engineer who, becoming a self-conscious audience, reads and re-interprets the notes made earlier. The manual represents this new effect to the technicians. To them, it is a presence that stands for the absent engineer. The effect of the technicians' interpretation is represented to the supervising engineer through the actions they perform. These "physical signs" are observed and reported by the supervising engineer, who in leaving a record of the project invites other readers (managers or engineers working on similar projects) to extend the semiotic chain.

What are the potential practical gains made by giving this theorized description of the communication situation? The model makes some important suggestions about the relationships and actions that compose the overall situation, suggestions that, if heeded, could improve the quality of technical communication. Three such provisions come immediately to mind:

1. Not only words, but also people, thoughts, things, and actions require interpretation; nothing is transparent, or automatically clear, but everything must be made meaningful by an interpreting mind. The engineer evaluating the project, for example, can't just report what was done, but must also say

how the actions did or did not accurately represent the original objectives and how, given the way things turned out, those objectives did or did not effectively interpret the needs. *The writer must re-create the semiotic chain so that others may extend it.*

2. The model also suggests that texts not only relay information but also "stand for" people. In the absence of the engineer, the technicians turn to the manual. To be effective, the manual must be complete enough and clear enough to substitute for the actual presence of the engineer. This understanding lies behind the many recent attempts to insert a more effective persona into technical documents, particularly manuals [10, 11].

3. Since, in each conversion of one sign to another, the medium of representation changes—from thought to text, from text to action—the substance of what is communicated is likely also to change. Needs cannot be fully comprehended in a single set of objects; in one sense, the objectives *reduce* the needs to the verbal medium. But, in another sense, the needs are *constructed,* given a rational form for the first time. Likewise, a person's thoughts are not exactly rendered in a written text, but are rather given concrete form and are fully developed for the first time. Nor will actions correspond point for point with a text on which they are based; when workers do a job, they will omit some recommended actions, they will embellish some others, and they will add some of their own. *In each exchange, something is lost and something is gained*—an observation which we call *the principle of rhetorical compensation.* The project engineer, for example, loses control over the details of action by allowing the manual to substitute for personal direction when the technicians do the job, but the engineer gains the freedom to work on other projects by using the manual to extend personal presence. The manual is, however, a weak substitute for the actual presence of a knowledgable supervisor because it cannot adjust to contingencies as they develop in the actions of the technicians. The author could not have foreseen every contingency. Moreover, the technicians are likely to improve upon the method given in the manual, and, being absent, the engineer will not be able to incorporate these improvements into future projects. The additional presence of the evaluating engineer compensates for the project engineer's absence; the evaluation report (substituting for the presence of the evaluator and once again adding to the accretion of new subject matter) may eventually serve as the basis for a revision of the manual.

To arrive at a deeper understanding of semiosis and the practical consequences of viewing communicative relations in this way, let's look at each element in the semiotic relation—the object, the interpretant, and the representamen—one at a time.

THE OBJECT

When simple signs—triadic relationships among objects, representamen, and interpretants—form into a complex chain of semiosis, such as the one describing the engineering project, the chain remains open at both ends. We can trace its starting point backward or forward infinitely. Every sign presupposes an earlier sign, and every new sign produced is open to reinterpretation and reformulation.

Every object has thus been conditioned by previous semiotic chains, especially by social and historical forces active in the language community of the makers and users of the chains. In Peircean semiotics, *the object is not necessarily a physical thing* (despite the unfortunate fact that people use the words *object* and *thing* interchangeably), nor is it merely a mental state. It could be either—and it could be much more: a word, an action, or even a total person.

The object's significance lies not in what it *is*, but in what it *does*. The emphasis on function and action—on the question *how?* and the verb *do*—is characteristic of the pragmatic strain of American philosophy, which begins with Peirce.

Peirce's definition of the object is close to that of the word *objective*—the starting point of action that, within itself, contains a preferred result. We still use the word *object* in this way when we say sentences like "The object (or objective) of my work is to build a better mousetrap." The subtlety of defining the object as an objective is not to be overestimated. Thinking this way allows us to resolve the long-standing conflict between physicalist and mentalist models of perception and communication.

The Physicalist and the Mentalist View of the Object: It and I

This concept of the endless semiotic series, the chain open in both directions, suggests that models of technical writing that locate the origin of communication in either the things of the physical world or in the mind of the communicator have oversimplified the relation of signs. Either model—a thing-centered (physicalist) model or a mind-centered (mentalist) model—presupposes a *transcendental object*, a point of origin for the semiotic chain [1, p. 325]. In the physicalist model, this object is the bedrock reality of physical nature. In the mentalist model, it is the psychological reality of the human mind or self, as theorized most clearly by Descartes, who said, "I think, therefore I am" (and thereby implied a strong comparison between the human mind and the Jewish God of Exodus 3.14, known simply as "I AM").

As Peirce recognized, the dualism of mentalism and physicalism is written into the very structure of our language. We talk about the relationship of mental and physical reality by using sentences. And English sentences, like those of most Indo-European languages, take the basic form of *subject-verb-object*. Both physicalism and mentalism freeze the world in this grammatical relation. The difference lies in which element of the grammar each camp recognizes as primary.

Mentalists choose the subject and generally give it a human character. *I think; therefore I am* comes to mean that *I think the world and even myself into existence.* I therefore privilege the human mind over any objective reality. The semiotic chain starts with *me.*

Physicalists, on the other hand, privilege the object as defined in the traditional sentence grammar and view the world as acted upon rather than as active. No wonder that scientific writing prefers the passive voice: *The world is constituted* rather than *I constitute the world.* The semiotic chain starts with *it,* the world. Reality is given. Even we ourselves may be objects, not subjects. We may conceive of ourselves as perceiving the world much in the manner of a machine with an optical scanner and word processing capabilities.

Throughout this book, we will argue that both mentalism and physicalism are inadequate as starting points for modelling communication. According to the theory of the triadic sign, both mentalists and physicalists are grasping only one side of a more complex, but still fundamental relationship. In later chapters, we will consider in more detail some of the limitations of I-based rhetoric (the problem of the "persona" discussed in Chapter 6) and it-based rhetoric (the problem of the "thingish" style, discussed in Chapter 7). For now, we only want to consider their shared weakness. This weakness amounts to a failure to recognize the triadic nature of semiosis.

In representing a true alternative to physicalism and mentalism, Peirce's triadic model presents an opportunity to overcome some of the deepest conflicts in the recent theory and practice of technical composition, for technical communicators often get stuck in either a physicalist or a mentalist view of their work.

Teachers and practitioners oriented primarily toward the physicalist perspective, usually those with an engineering background, insist that the main aim of technical writing is to give a clear and accurate rendering of a dominant physical reality. They believe that technical writing is primarily descriptive. When they write or use textbooks, they focus on exercises like the description of a mechanism and advocate the use of a plain, but relatively impersonal style that gets the job done with no frills.

Their counterparts, the mentalists, are the new wave of humanists who have entered the field of technical communication since the late 1970s. These teachers and scholars insist that the best technical writing is people-conscious, cognizant of the needs of the audience and demonstrative in asserting the presence of the author. They have bolstered technical communication theory by introducing techniques and maxims from rhetoric, the classical art of persuasion. When they write or use textbooks, they focus on procedures for "invention" of texts and advocate the use of a personalist style that engages and encourages the audience.

In fact, the lines are not so clearly drawn. Most technical communicators, when they stop to reflect, can see merits in both sides of the physicalist/mentalist conflict. Only the most dyed-in-the-wool mentalists would deny some measure of "reality" to the world described by the physicalists. And few physicalists would refuse to recognize that their view of the world is greatly altered by language and custom: a Trobriand Islander sees a world very different from that seen by the anthropologist who visits the native village.

Most teachers and writers thus already tend to be pragmatists in their alternate use of physicalist and mentalist communication techniques. They know that, at various moments, either approach can be useful. By incorporating both the physicalist and the mentalist viewpoints within a single semiotic chain, Peirce's theory shows the way to resolving the theoretical conflict in a more systematic way.

Dyads and Triads: A Crucial Difference

Both physicalism and mentalism are essentially *dyadic*. Physicalists and mentalists both view the world in twos, as a relationship of a subject and an object. So where does the text fit in? What about the medium that intervenes between subject and object? Is it not active as well? Instead of saying "*I* (grammatical subject) do something to *it* (grammatical object)," can we not say, even in our limited grammar, "This *manual* helps *me* do something to *it*"? Peirce goes even further than this.

Human beings, he argues, must use signs to mediate their understanding of both subjects and objects. In addition, however, he claims that subjects and objects, both *I* and *it*, are not only things to be mediated by signs but are themselves signs. His concept of the "object" in the chain of semiosis should therefore not be confused with the "object" in the traditional grammatical relation or the notion of an "object" as a thing in the physicalist sense. It is much more than either of these.

Peirce defined the object as potentially active *and* passive, productive and consumable, a state implicit in the concept represented by the noun *objective*. A number of good how-to books on technical writing already advocate document planning on the model of management-by-objectives—Cunningham and Cohen's book on creating technical manuals, for example [12]. Consciously or unconsciously, these writers ground their work in Peircean pragmatism.

An objective as conceived in the notion of management-by-objectives is a kind of goal that shapes and motivates both action and reflection in project planning, implementation, and evaluation. Likewise, the Peircean object determines the meaning of its interpretant (in a relation that would please a physicalist), but it also is conditioned by that interpretant (in a relation that would please a mentalist).

In our example of the engineering project, the need to be met by the engineer's actions is an object in this sense. It is active in the sense that it shapes the whole project, but passive in the sense that it is interpreted and given a written form by the project engineer. If the need was to get supplies and people from one side of a river to another, for instance, the engineer's options for action are clearly limited by the nature of the objective itself (the object), but the decision to build a bridge or develop a ferrying service would depend entirely upon the engineer's judgment (the interpretant). That judgment—to build a bridge to accomplish the objective—in turn becomes a new objective replacing the old one. Now, as a new object, it must be interpreted in diagrams, documents, and ultimately in action.

By treating the structure of semiosis in diagrams rather than in sentences, Peirce teaches us to capture this concept of the object as a Janus-faced entity, a piece of the writing process that looks backward toward physical reality and forward toward the mind that interprets (or even *constructs*) that reality. One of the reasons we need to see the whole chain of triads before we can firmly grasp the meaning of the simple sign is that we are tempted to view the simple sign as a closed system that originates with some primordial, "real" object, whether that object is the "real world" of physicalism or the "thinking subject" of mentalism [13].[5]

The chains of triadic signs in our model of the communication process, on the other hand, suggest that texts are transformed at each transaction and that texts and actions are interchangeable signs. An author interprets physical reality or thoughts about that reality (usually both) and represents them in speech or writing. An audience then interprets the author's representation and re-represents it in further speech or writing—or action. The text (itself an object) is always "completed" by the reader [14].[6]

What kind of reality, then, is set forth in the Peircean scheme, and how does it differ from the physicalist or mentalist reality it replaces? What kinds of actions are possible through semiosis, and how may we move from the endless chain of semiotic relations back to the "real world"?

Deconstructionists may deny that this move is possible, suggesting that the "textualized world" only arises from and refers to other texts. Writing always originates and terminates with other writing. But the pragmatist Peirce never doubted the possibility of building texts that may have an empirical referent (a base in physical reality) and that may be re-converted into action. But neither did he deny the problem of how to convert textual action to physical action, and vice-versa.

Instead, Peirce theorized two kinds of action that affect human existence—*dynamic action* and *semiotic action*. Whereas semiosis is triadic, dynamic action is dyadic. In dynamic action, an object exerts a force upon another object.

Human *bodies* daily suffer the blows of nature—accident, disease, aging—and are susceptible as well to violence from other human beings. All of these are originally dyadic actions, forces acting upon objects, objects in brute motion. But, according to Peirce, human *minds* are less subject to dyadic action. They convert dynamic (dyadic) actions to triadic signs in the process of semiosis.

An Illustration of Dynamic and Semiotic Action

A geologist who finds a rock and interprets a set of marks upon it engages forces that were once dynamic and dyadic, but which, for the purposes of science, he or she renders as semiotic and triadic. In the diagram in Figure 6, the arrow

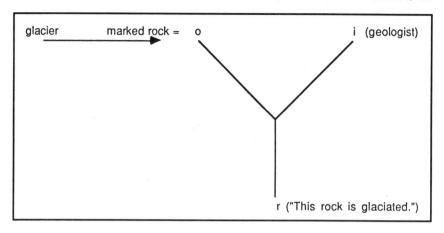

Figure 6. Dynamic action converted to semiosis.

represents the action of a dynamic, dyadic relation in nature: A glacier leaves scratch marks on hard rock. The accompanying triad shows that, to understand what caused the scratches on the rock, the geologist interprets the marks, deducing that a passing glacier has left its imprint on the rock, thus creating a presence (an interpretive statement) that stands for an absence (the long-melted glacier).

The rock, the object of investigation, is not, as a physicalist would argue and as this fragment of a true semiotic chain would suggest, the point of origin for the study. The selection of the rock, even the geologist's notice of it, and most certainly the interpretation of the scratches upon it are conditioned by what the philosopher of science Thomas Kuhn has called "normal science" in his celebrated book, *The Structure of Scientific Revolutions* [15].

The practice of normal science involves a standard set of established theories and problems that geologists agree will concern them. Normal science is profoundly semiotic and therefore triadic. The geologist brings the glaciated rock into the communal process of geological semiosis. The dynamic action of the glacier upon the rock is itself external to the semiosis.

Perhaps, then, it would be better to bracket the dynamic action in order to show its true place in the semiosis of the geological interpretive community, as in Figure 7.

In Peircian pragmatism, the world of dynamic action, though certainly real, is external to the world of semiosis. "We will define the real," Peirce writes, "as that whose characters are independent of what anybody may think them to be" [5, vol. 5, p. 405]. Anything can be a sign, so long as it is interpreted as such [1, p. 330]. But, as soon as a person treats dynamic action as a sign, it "becomes" rational, triadic, semiotic, and thereby subject to interpretive and representational processes.

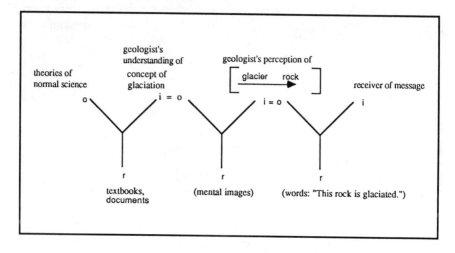

Figure 7. Semiosis of "normal science."

Semiosis is Both Internal (Subjective) and Communal (Intersubjective)

Semiosis is thus internal, a psychological dialog within the mind of a single thinking subject, or it is intersubjective, an interchange between two thinking subjects or among several members of an interpretive community. In certain writings, Peirce was skeptical about the effects of semiosis on human behavior and about our ability to have any effect on the outer world of reality. The difference between the internal or intersubjective world of semiosis and the external world of dyadic force, he wrote at one point, is this [5, vol. 5, p. 475]:

> ... the Inner World ... exerts a comparatively slight compulsion upon us, though we can change it greatly, creating and destroying existent objects in it; while the ... Outer World ... is full of irresistible compulsion for us, and we cannot modify it in the least, except by one particular kind of effort, muscular effort, and but very slightly even in that way.

The interaction of these two worlds therefore mainly consists in "a direct action of the outer world upon the inner and an indirect action of the inner world upon the outer" [5, vol. 5, p. 493]. The effect of the inner world upon the outer is indirect because it is mediated by signs. Only through the control of the signifying medium can the "real world" come under rational control. Pure dynamic action is dyadic, but *conduct* or *rational action* is triadic [5, vol. 1, p. 337].

This distinction leads us back to our diagram of the triadic chain in Figure 5, reproduced in Figure 8.

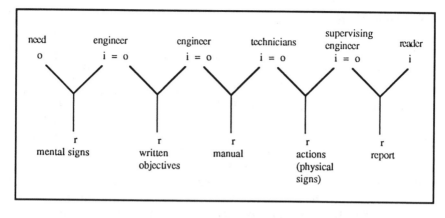

Figure 8. Semiosis in an engineering project
(reproduced from Figure 5).

The work of the technicians is muscular work that has an effect upon the physical world not qualitatively unlike the work of the glacier upon the rock. But their work, unlike the work of the glacier, is not dyadic; it is triadic. It is ideally guided by the engineer's discourse and is thereby not mere brute force but rather semiotically mediated force, the product of representation and interpretation. And this rational action is likely to instigate further semiosis. Human work, unlike the work of a glacier, is open to the possibility of rational control and the corrective influence of mental and verbal constructs.

Writers never *merely* represent the world, as physicalist rhetorics would suggest. They represent the world so that they can do something, so that they can *intervene* in reality [16].[7] How they intervene is largely an effect of how they interpret their own representations of reality and other people's representations. Now is thus the time to consider how, in the Peircean scheme, interpretation affects the way we act.

THE INTERPRETANT

"When we think," Peirce writes, "then we ourselves, as we are at that moment, appear as a sign" [5, vol. 5, p. 283]. The self is a sign, an *image* we project as we reflect upon ourselves or present our ideas to others. As signs, we take our places in the semiotic chain and must be determined by other signs, for "there is no exception . . . to the law that every thought-sign is translated or interpreted in a subsequent one, unless it be that all thought comes to an end in death" [5, vol. 5, p. 284]. Just as Peirce's notion of the object transforms physicalist thinking, his concept of the interpretant alters mentalist notions of the self.

The Self Is a Sign

Human beings, like glaciers and rocks, have an existence in the world of dynamic action. Our thoughts have relatively little effect on that world. Maybe we can predict an earthquake, but we can't stop it.

However, we do have great power to affect our *selves, our written texts, and ultimately our own actions.* As soon as a person chooses to enter the semiosis of geology or engineering, that individual is re-created as a geologist or an engineer. The identity of the new geologist will be largely determined by the behavioral, intellectual, and verbal norms of professional affiliation.

Just as individuals change, the communities of normal science, which are themselves possessed of an image, will also change. The possibility of "revolutionary science" [15]—the ability of interpretive communities to transform themselves by transforming the theories or "paradigms" that guide their work—remains open for two reasons.

First, revolutionary science is possible because dynamic action may radically influence a community's thinking. New events in the external world—the AIDS epidemic, for example, or an "ozone hole" over the South Pole—may force themselves upon the community's attention so strongly that people must invent new interpretations to account for them. Likewise, new technologies may permit scientists to see facts in a way that undermines former theories.

Second, the possibility of revolutionary science remains open because of people's personal and social complexity. Everyone is constituted by more than one community or subcommunity at any given moment. Thus Kenneth Pike can observe that "the hopeless attempt to eliminate the observer in favor of a scientific detachment or objectivity is dismal" [7, p. xii]. "Scientific detachment" is the power of a community to control an interpreter to the point that the effects (interpretants) of all other communities are obliterated. "Objectivity" is the power of an interpreter to focus only on the objects deemed theoretically appropriate by the community attempting to limit interpretants. As Pike's comment suggests, we can be neither perfect embodiments of community standards nor infallible observational instruments built according to the specifications of that community.

Peirce's version of pragmatism preserves, but at the same time qualifies and demystifies, the acting, interpreting human subject—the hallmark of the mentalist perspective. The self becomes one sign among many. A key text on this point comes from Peirce's letter to Lady Welby of 23 December 1908, which we quoted in part earlier in this chapter [6, pp. 80-81]:

I define a Sign as anything which is so determined by something else, called its Object, and so determines an effect upon a person, which effect I call its Interpretant, that the latter is thereby mediately determined by the former. My insertion of "upon a person" is a sop to Cerberus [the mythic three-headed hound that guards the gates of Hell], because I despair of making my own broader conception understood.

Peirce recognized that "it is not merely a fact of human psychology, but a necessity of Logic, that every logical evolution of thought should be dialogic," that semiosis—the process of this "logical evolution"—requires at least two "quasi-minds," a "quasi-utterer" and a "quasi-interpreter" [5, vol. 4, p. 551]. What does Peirce mean when he uses "interpretant" rather than "interpreter," "quasi-utterer" rather than just plain "utterer," and "quasi-interpreter" rather than "interpreter." As the Peircean scholar Max Fisch notes, Peirce seems to have felt uncomfortable "lapsing from sign-talk into psych talk—from semeiotic[8] to psychology" [1, pp. 3423-343]. Why?

"By semiosis," Peirce explained, "I mean an action, an influence, which is, or involves, a cooperation of three subjects, such as a sign, its object, and its interpretant . . " [5, vol. 5, p. 484]. But, as the Italian semioticist Umberto Eco has noted: "the 'subjects' of Peirce's 'semiosis' are not human subjects but rather three abstract semiotic entities, the dialectic between which is not affected by concrete communicative behavior." Eco continues with a qualification: "I do not deny that Peirce also thought of the interpretant (which was another sign translating and explaining the first one, and so on *ad infinitum*) as a psychological event in the mind of a possible interpreter; I only maintain that it is possible to interpret Peirce's definition in a non-anthropomorphic way" [17, p. 15].

To define the sign relation anthropomorphically—in terms of personal human psychology—means to weaken its power of generalization. Hence Peirce was concerned about having offered a "sop to Cerberus." In order to get us readers beyond the gate of semiotic hell, into the difficult regions of deep theory, Peirce bribes the guards, our defense mechanisms, that protect our minds from unwanted (or unneeded) complexities. He lets us think about the interpretant as if it were a human being, an interpreter.

Peirce defined the interpretant as the "significate [sic] outcome of a sign" [5, vol. 5, p. 473]. Whatever results from the interchange of the object with its representamen—the message with its medium—is the interpretant. In this sense, the interpretant is above all an altered message, a new version of the object.

The process of semiosis thus changes thoughts, words, and deeds—selves, texts, and objects—by re-representing them. Eco claims emphatically, *"The interpretant is not the interpreter* (even if a confusion of this type occasionally arises in Peirce)" [17, p. 68]. Perhaps it is better to say that the interpretant is not *only* the interpreter. As Eco correctly argues, "the idea of the interpretant frightened many scholars who proceeded to exorcise it by misunderstanding it (interpretant = interpreter or receiver of the message)" [17, pp. 69-70]. The idea frightened scholars—or rather *confused* scholars whose understanding was above all psychologically oriented around an active human subject—because it threatened the mentalism of independent and glorified human selfhood, by suggesting that the self is one sign among many.[9]

But, in his effort to exorcize the psychologism of Peirce's followers (especially behaviorists like the influential Charles Morris), Eco goes too far in restricting the meaning of the interpretant. For him, the interpretant is simply a linguistic principle, a definition. Peirce says that the interpretant "need not be of a mental mode of being" [5, vol. 5, p. 473]. Nor need it be exclusively linguistic in nature, as Eco seems to imply at times. We could, as we have suggested, place three kinds of entities in the position of interpretant:

- a person, or psychological interpretant
- a text, or textual interpretant
- an action, or physical interpretant

As it turns out, these are precisely the three things in this dynamic world that we have some degree of control over—our selves, our texts, and our actions.

Pragmatic Interpretation Focuses on Actions

If we think about the interpretant as an "outcome," we can see that the action (the physical interpretant) and not the person (the psychological interpretant) fits the definition best.

This emphasis on purposeful action is at the heart of pragmatism as it has evolved since Peirce, especially in the work of John Dewey and George Herbert Mead. In his classic essay "How to Make Our Ideas Clear," Peirce gives this rule for attaining the highest degree of clarity: "Consider what effects, which might conceivably have practical bearings, we conceive our conception to have. Then, our conception of these effects is the whole of our conception of the object" [9, p. 37]. Peirce is, in this way of thinking, the unacknowledged father of management-by-objectives. "The whole function of the mind," he wrote in an unpublished lecture, "is to make a sign interpret itself in another sign and *ultimately in an action*" [9, p. 66; italics added].

We can determine the clarity of a sign, therefore, neither by some intrinsic quality of the language in which it may be expressed (the plainness of the language, for example, its lack of jargon and its active verbs) nor by its faithfulness to reality or its truth value, but rather by its ability to project a desirable outcome.[10] A technician does not read a manual to discover the truth, but to perform an action in the right way. A supervisor does not read a report only to be enlightened, but to decide how to award credit and to plan future actions.

And what of scientific readers? They are presented with documents that the rhetorical theorist Walter Beale [19] has seen fit to designate as the "contemplative" pole of technical communication set at some distance from manuals and proposals at the "instrumental" pole. But, in fact, scientific documents also enable actions; a scientist reads the literature *in order to do further research,* not merely to gain "contemplative" understanding that will never be put to use. (The very word "under-standing" suggests a preparation for action, a solid base from which to work.)

If actions, real physical outcomes, are the pragmatic interpretant par excellence, what about the other types of interpretants we have described—persons and texts? In our action-oriented rhetoric, these can be conceived as intermediate steps to actions. The image of the self, for example, provides a step to action in this sense: Before acting in a rational manner, we have to be prepared. We must experience an alteration of consciousness, must (in the language of semiotics) be *reconstituted* as a new self.

The giving of knowledge, the process of education or initiation into a community of discourse, prepares people for action. If studying glaciation, the geologist must learn the methods of observation and experimentation certified in the geological community before acting like a geologist. Students preparing to write informative reports and manuals must first recognize that they can no longer write exclusively for the teacher. They cannot write from a position of lesser knowledge to a position of greater knowledge, but quite the opposite: They must write as if they know more than the reader. They must themselves become patient, careful teachers, since the professional technical writer always writes for a reader who needs (or wants) to know the information the writer possesses. Students must be reconstituted as *knowing* authors before they can be successful even as quasi-professional writers.

Like a successful image of the self (a *persona*, or mask), an effective text also provides a step toward action. It *substitutes for a person or an action*. Note-writing provides a simple example of this act of substitution. A person writes a note to remember to do something, maybe take out the garbage, knowing full well that, when the time arrives for garbage pick-up, other thoughts may interfere with the action. The term people apply to this process is *to remind*. To re-mind means to reconstitute the state of mind they had when they first conceived of the desired action (taking out the garbage). The text of the note ensures the proper action at the proper time.

By interpreting self, thoughts, and objectives in written texts, writers attain new levels of mobility and effectiveness. They can use their minds for other jobs and can forget about planned actions until they review their notes for daily activities. More important, they can leave the note for someone else, who can read and act in their stead. The teen-aged son or daughter can carry out the garbage when the parents go out of town.

Of the three types of interpretants, the written text most readily fits the ancient definition of a sign as *a presence which stands for an absence*.

Refining the Model

Let us now reconsider the chain of triads that we used to diagram the semiosis of a hypothetical engineering project presented in Figure 5 (reproduced in Figure 9 for purposes of comparison).

Now, with our de-psychologized version of the interpretant in place, let's look at an expanded and revised version of this diagram in Figure 10.

Both figures show the semiotic chain developing from the perceived need to the final report. But the perspective shifts when we have a richer view of the interpretant as person, text, or action. The version in Figure 9 is distinctly mentalist or humanistic. It offers the sop to Cerberus by favoring the human agent as the exclusive interpretant, while the version in Figure 10 shows a radicalized view of text and physical behavior. Texts and actions now become active agents in the same way that selves are active agents.[11]

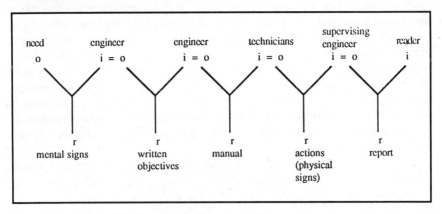

Figure 9. Semiosis in an engineering project
(reproduced from 5).

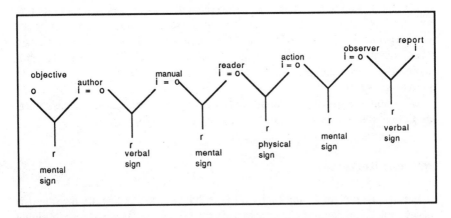

Figure 10. Expanded diagram of the semiosis of an engineering project.

The chief advantage of our model may be realized as *socializing the sign process—taking it out of the individual mind and representing its relations among minds,* as in our diagrams of the work of the engineer and geologist. Peirce may have seen this as yet another concession, another sop to Cerberus. As the philosopher Richard Bernstein has remarked, Peirce was "unmoved and even suspicious" of the "practical turn" of pragmatism that led to formulations like Dewey's "conception of a guide to social reconstruction" [20, p. 173]. Our rendering of Peirce's semiotic is likewise a taint on the purity of the Peircean "pure rhetoric." But, though Peirce was reluctant to admit his patrimony, he gave us the "concept of self-controlled conduct" that, in the words of Bernstein, "provides the mediating link between the traditional dichotomies of theory and practice, thought and action" [20, p. 189].

Because both objects and interpretants can exist in physical, mental, or textual states, an interpretant in one sign relation is able to become an object in another, as the signs of equivalence (=) in our diagrams indicate. Such is the transformation implicit in the semiotic series: Interpretation gives rise to re-interpretation. The medium through which this interchange occurs is the representamen, which we will now consider.

THE REPRESENTAMEN

Peirce defined *represent* as "To stand for, that is, to be in such a relation [to some entity] that for certain purposes it [another entity] is treated by some mind as if it were that other" [5, vol. 2, p. 273]. A representamen "stands in such a genuine triadic relation to . . . its *Object,* as to be capable of determining its *Interpretant* [5, vol. 2, p. 274].

Tossing overboard the physicalist view about the nature of sign and referent, the pragmatist refuses to say that to represent means to reflect. The representamen has a much more active role; it functions in an attempt to *render the object and the interpretant equivalent.* In line with this view, the modern critic and philosopher Kenneth Burke made a deep contribution to the theory of rhetoric by treating *identification* as the central concept of his *Rhetoric of Motives* [21].

In Peircean theory, the representamen is the medium through which the object is identified with the interpretant, and which, through a further medium, may be treated as an object itself. This is not a simple identification, in which one thing is made the exact equivalent of another. Rather it is an identification that is more like a metaphor, in which one thing, though different from another, is said to be equivalent and is transformed in the process by the work of representation and interpretation. In saying "love is a rose," the poet alters the reader's understanding of love and roses.

This medium therefore radically affects the nature of the interpretant. Though somewhat simplistic and reductive, the slogan of the provocative analyst of mass communication Marshall McLuhan—*the medium is the message*—has become justly celebrated.[12]

As we shift media in a presentation, we deeply affect how the reader receives the message. A picture of an object will produce an effect—an interpretant—very different from a word describing that object. In the next chapter, we will refer to this effect as the *principle of complementarity,* which suggests that a shift in the medium of representation (the representamen) cues the reader to a shift in meaning (in the relation of the object to the interpretant). No part of any sign can be changed without changing the total effect of the sign. No alteration in a text is ever innocent of effects upon the content presented. Presentation is always representation.

The classification of various media therefore becomes an important activity for semiotics. For our model of technical communication to be complete, we will have to develop a system of classifying and predicting the behavior of the different signs or media, the effects of representamen on their related objects and interpretants.

Peirce himself attempted to deal with the difference between symbols and images in one of his many classifications of sign types—the famous trichotomy of *icon, index, and symbol,* to which we turn in Chapter 3. We will show how the various features of technical texts correspond to these three sign types. In Chapters 4, 5, and 6, we will consider another approach to classifying the representamen or media of technical communication—as the genres or social codes of technical documents. Even the communities of discourse that we take up in Chapters 7, 8, 9, and 10 may be considered as media through which information must pass on its way to being represented and interpreted by communicators.

Turning to Chapter 3, it will be helpful to remember a few key points from this chapter:

1. The concept of the sign is useful in technical communication theory because it is flexible enough to accommodate many different kinds of textual elements—words, pictures, graphics, even persons, physical objects, and actions.
2. Communication, thought, and rational action depend upon the presence of signs.
3. A sign is a presence that stands for an absence, a triadic relationship in which a representamen (medium) is used to create a new relationship (usually a transforming identification) between an object and an interpretant.
4. The process of converting thought to text and text to action (and vice versa) is best described as an open semiotic chain. No object is transcendental and no interpretant is the final word on the subject. One sign continuously gives rise to another sign.
5. Representation in signs requires interpretation. Nothing is ever transparent or literal.

6. Technology is possible because semiosis allows us to control and shape the actions of other people and (to a lesser extent) the actions of nature.

7. The pragmatic model that treats action as the ultimate outcome of writing and reading is best suited as a theory for technical communication.

NOTES

[1]The difficulty of distinguishing the *sign* as a whole relationship from *semiosis* as the process by which signs function is the kind of problem that has led the Italian semioticist Umberto Eco to argue that we should scrap the notion of the sign as a thing and totally replace it with the idea of *sign-function*. The idea that there are no signs but only semiotic functions is an appealing possibility, one which in Chapter 3 we will adopt fully. For now, however, it will be easier for us to provide an initial description of Peirce's theory of communication if we avoid this subtlety and keep the concept of sign-as-thing intact. In our culture's dominant (and essentially positivist) way of thinking, it is easier to gain an initial understanding of a thing than it is to comprehend what is inevitably a complex and diffuse process.

[2]A step further in the argument would take us to the point of radical idealism—to the claim that the object only comes into existence in the presence of a representamen and an interpretant—but, as we shall show later in the chapter, Peirce was too much of a realist to make this claim.

[3]The analogy, we hope, would have appealed to Peirce, who was a trained chemist and physicist. In speaking of the "valency" of signs, he drew a similar analogy. The triadic diagrams rightly remind many readers of the molecular chains used in the diagrams of organic chemistry.

[4]The diagram, though greatly expanded, is nevertheless a radical simplification that in many ways destroys the subtlety of Peirce's concept of the chain of triadic relationships. Above all, it still substitutes the idea of interpreter for the concept of interpretant (the *effect* upon an interpreter). We shall see later in this chapter that the "psychologizing" of the interpretant was a worry of Peirce's, and that the original notion allowed that even a written document or a word—since it interprets other words and represents an effect upon a mind—could be considered an interpretant.

[5]Following Kant, the deconstructionist philosopher Jacques Derrida identifies this primordial thinking subject as the "transcendental ego." In his now famous critique of the closed structure, Derrida extends the tradition of semiotics that has flourished in France. He gives a Peircean flavor to the dyadic formula of the father of European semiotics, Ferdinand Saussure. For Saussure, the sign was composed of a *signified* and a *signifier*. Derrida treats the signified as more or less the equivalent of Peirce's object and the signifier as the equivalent of Peirce's interpretant. Adopting the Peircean notion of the endless series of sign relations, Derrida denies the existence of a primordial "transcendental signified." He cites the kinship of Peirce's view with his own: "Peirce goes very far in the direction I have called the de-construction of the transcendental signified, which, at one time or another, would place a reassuring end to the reference from sign to sign" [13, p. 49]. Peirce admits that the kinds of signs that permit communication—which he calls *symbols*—may "come into being by development out of other [simpler] signs, particularly from icons," which may have a more basic relation to the real world. Nevertheless it is only "out of symbols that a new symbol can grow" [5, vol. 2, p. 302]. Derrida agrees with Peirce that the roots in

simpler phenomena "must not compromise the structural originality of the field of symbols." Thus: "No ground of nonsignification—understood as insignificance or an intuition of a present truth—stretches out to give it foundation under the play and the coming into being of signs" [13, p. 48]. Everything is a sign once it enters consciousness, though there are different kinds of signs with different kinds of relationships to other realities—such as the two mentioned here, the icon and the symbol. We can determine the differences among signs by the kind of relationship the representamen has to the object, as we shall show in the section on representation later in this chapter and in Chapter 3.

[6]Karen Burke LeFevre has argued this point forcefully in her work on the social nature of the writing process. In claiming that rhetorical invention—the creation of texts—is a cooperative project between writers and their readers, LeFevre writes, "I rely on Hannah Arendt's discussion of 'action,' which in turn derives from the Greek and Latin double meaning of the word. Action, Arendt says, was previously regarded as having two parts: 'the beginning made by a single person and the achievement in which many join by 'bearing' and 'finishing' the enterprise, by seeing it through.' The word has since lost this double meaning, . . . with action being split into the role of giving commands and the role of executing them. The one who initiates commands now appears to be isolated from others. We may regard such a person as independent and powerful in her apparent isolation, but in fact, Arendt insists, such a person is not isolated at all; the potential for power requires the presence of others, and the achievement of action requires that others execute and thus complete the action" [14, p. 38]. In this argument, and in her comparison of rhetorical invention to invention in the technological sense, LeFevre's theoretical project shares much with our own.

[7]In an excellent summation of this pragmatic point, the philosopher of science Ian Hacking writes: "Science is said to have two aims: theory and experiment. Theories try to say how the world is. Experiment and subsequent technology change the world. We represent and we intervene. We represent in order to intervene, and we intervene in light of representations. . . . The final arbitrator in philosophy is not how we think but what we do" [16, p. 31].

[8]Peirce preferred this spelling and the pronunciation SEE-MY-OH-TIC, and Fisch, following Peirce, argues that this spelling and pronunciation are more etymologically correct than the more popular form, "semiotics," pronounced Semi-AH-TICS. In this book, we use "semiotic" to refer to a single system, much in the same way we use "rhetoric"—the rhetoric of Aristotle and the semiotic of Peirce. We occasionally use "semiotics" as well as "rhetorics" as a plural form when we speak of plural semiotic or rhetorical systems. (Compare "ethics" and "poetics.")

[9]In this vein, Peirce joins other nineteenth-century philosophers—above all Darwin and Freud—in chastening the exalted image of humankind received from the Enlightenment.

[10]The contemporary pragmatist, Richard Rorty, in a direct challenge to physicalist logic and rhetoric, argues that "the notion of 'accurate representation' is simply an automatic and empty compliment which we pay to those beliefs which are successful in helping us do what we want to do" [18, p. 10].

[11]There is at least one remaining problem with Figure 10: It reduces the concept of the representamen to an abstraction. A deconstructionist or a nominalist would argue that therefore the representamen has become "metaphysical" and may be eliminated from the model, leaving us with a more or less dyadic, Saussurean picture in which the object corresponds to the signified and the interpretant to the signifier; this is essentially what

happens to the Peircean scheme in the hands of Derrida and Eco. However, we feel it is necessary to retain the representamen as a reality, a designation of the "medium" of interpretation. It is this medium that is the motive force in causing the change from object to representamen, which would otherwise be a simple chain of identities. In Chapters 3 and 4, we will give terms that replace the makeshift abstractions of "mental sign," "physical sign," and "verbal sign" by introducing the concept of sign-types (icon, index, and symbol) and genres (codes).

[12]McLuhan's identification of the medium and the message is a nominalistic reduction similar to that of Derrida. Both of these analysts seem to recognize the parts of the triadic sign, but both create the conditions for at least one of those parts to be eliminated as metaphysical or unreal, thus rendering the triadic relation as dyadic. McLuhan's saying suggests that we can treat the medium and the message as a single entity, for example. But could McLuhan have been speaking metaphorically? In one sense, we could say that his semiotic is much the same as Peirce's: The medium is the message in the same way that the object is the interpretant. Because of their association in a semiotic system, the medium and the message become linked inextricably to one another. To change one would mean to change the other.

REFERENCES

1. M. Fisch, *Peirce, Semiotic, and Pragmatism: Essays by Max Fisch*, K. L. Ketner and C. J. W. Kloesel (eds.), Indiana University Press, Bloomington, 1986.
2. V. N. Volosinov, *Marxism and the Philosophy of Language*, L. Matejka and I. R. Titunik (trans.), Seminar Press, New York, 1973.
3. J. R. Lyne, Rhetoric and Semiotic, *Quarterly Journal of Speech, 66*, pp. 155-168, 1980.
4. R. Scholes, *Semiotics and Interpretation*, Yale University Press, New Haven, 1982.
5. C. S. Peirce, *The Collected Papers of Charles Sanders Peirce*, 5 vols., C. Hartshorne and P. Weiss (eds.), Harvard University Press, Cambridge, Massachusetts, 1960.
6. C. S Peirce, *Semiotic and Significs: The Correspondence of Charles S. Peirce and Lady Victoria Welby*, C. S. Hardwick (ed.), Indiana University Press, 1977.
7. K. Pike, *Linguistic Concepts: An Introduction to Tagmemics*, University of Nebraska Press, Lincoln, 1982.
8. J. Ransdell, Some Leading Ideas of Peirce's Semiotic, *Semiotica, 19*, pp. 159-178, 1977.
9. C. S. Peirce, *Classical American Philosophy: Essential Readings and Interpretive Essays*, J. J. Stuhr (ed.), Oxford University Press, New York, 1987.
10. D. Bradford, The Persona in Microcomputer Documentation, *IEEE Transactions on Professional Communication, 27*, pp. 65-68, 1984.
11. B. Rubens, Personality in Computer Documentation: A Preference Study, *IEEE Transactions on Professional Communication, 29*, pp. 56-60, 1986.
12. D. Cunningham and G. Cohen, *Creating Technical Manuals*, McGraw-Hill, New York, 1984.
13. J. Derrida, *Of Grammatology*, G. Spivak (trans.), Johns Hopkins University Press, Baltimore, 1976.

14. K. B. LeFevre, *Invention as a Social Act,* Southern Illinois University Press, Carbondale, 1987.

15. T. S. Kuhn, *The Structure of Scientific Revolutions,* (2nd Edition), University of Chicago Press, Chicago, 1970.

16. I. Hacking, *Representing and Intervening: Introductory Topics in the Philosophy of Natural Science,* Cambridge University Press, Cambridge, 1983.

17. U. Eco, *A Theory of Semiotics,* Indiana University Press, Bloomington, 1979.

18. R. Rorty, *Philosophy and the Mirror of Nature,* Princeton University Press, Princeton, 1979.

19. W. Beale, *A Pragmatic Theory of Rhetoric,* Southern Illinois University Press, Carbondale, 1987.

20. R. Bernstein, *Praxis and Action: Contemporary Philosophies of Human Action,* University of Pennsylvania Press, Philadelphia, 1971.

21. K. Burke, *A Rhetoric of Motives,* University of California Press, Berkeley, 1950.

CHAPTER 3

Representation in Document Design

MANAGING REPRESENTATIONAL VARIETY

In Chapter 2, we suggested that one of the advantages of the semiotic model is that it allows us to describe and ultimately to manage a variety of textual features, both verbal and graphical. In this chapter, we will show how the theory of signs helps to account for the wealth of representational forms likely to appear in nearly any modern report, manual, or proposal.

Iconic-Mosaic Text and the Rise of Document Design

Even as early as the English Renaissance, when printing first caught on as a means of satisfying the growing middle-class readership, authors of how-to books used graphics, page design, and schematic presentation to break up the long blocks of prose associated with writing in the scholastic tradition [1]. In recent years, with the technical innovation of computer-based text production [2-5], and with the coordination of writers and artists with technical and marketing staff in integrated teams in industry [6], we have reached an advanced stage in the development of technical communication toward mixed media representation.

A technical writer can no longer be merely a wordsmith, but must create sentences, paragraphs, and other arrangements of words (notably lists) that complement and enhance the graphical features that tend to dominate the verbal text. Technical documents more and more typically are based upon what the media critic Marshall McLuhan has called *iconic-mosaic* design [7]. Instead of one great block of prose, chipped here and there by an occasional paragraph break or space for a title—the kind of text most of us learned to produce in school—technical communicators are encouraged to design documents with a stunning variety of textual features. The overall image rather than the flow of words is dominant, and the pieces of the page must subtly but effectively play off of one another to create a unified rhetorical impact.

Moreover, in the search for ever more readable and appealing texts, technical communicators have come under the influence of mass media page design. No doubt, this development will have both negative and positive effects upon communication. Negatively, we may witness an increase in "fluffy" presentations that sacrifice depth or even accuracy for powerful visual impact. More positively, the appeal of technical communication may increase, opening the access to technical information for an ever wider audience, including the growing numbers of nontechnical readers interested in technological products and services, such as business executives, office personnel, home users of personal computers, government decision-makers, and teachers. Such readers are likely to respond favorably to documents designed to remind them of their leisure reading in magazines and newspapers.

In the movement to accommodate nontechnical readers and to make use of the best combination of graphical and verbal effects in iconic-mosaic texts, "document design" has become a key issue in the practice of technical communication and in the education of technical communicators. Their work requires a keen visual awareness as well as a finely tuned verbal sensibility.

Avoiding the Traps of Bad Design

The technical communicator of our age must learn first of all to avoid the common traps of bad design. With desktop publishing, for example, the temptation is to overuse fancy typographical features, thereby creating semiotic "noise" in the text—bothersome distractions to good reading.

Another potential problem involves underrating the power of an image to draw the reader's attention. A designer can destroy the proper emphasis of a document by forcing the reader's attention to a relatively minor point that happened to have high potential for visual expression.

Good management of the complex visual dynamics of iconic-mosaic texts, whether laid out on a storyboard in a document-production group or created on a computer by a single author, thus requires special adjustments in textual analysis. Custom and experience will do part of the work. New writers will gradually learn to cope with the demands of iconic-mosaic text.

But only a solid grounding in theory can make sense of all the little rules and prohibitions that designers customarily draw upon. A theoretical comprehension of the function of textual elements enables designers at once to understand and to move beyond the appeal to "common sense," trial and error practice, and a loose sense of convention in document design (compare White [6, p. 5]). Just such a theoretical foundation can be derived from pragmatic semiotics, as we hope to show in this chapter.

As a start, writers would do well to take note of the powerful semiotic rule: *The addition of new signs or the conversion of one kind of sign to another cues the reader to a deep structural change, a change in approach or meaning.* Every sentence cannot become, willy nilly, a flow chart, and vice versa. Every document cannot be formatted like a user's manual. To replace one textual feature (a paragraph, for example) with another (a table, for example) means to create new interpretive needs in the audience while satisfying others. For, as we suggested in Chapter 2, *representation is always tied to interpretation.*

Verbal and Graphical Representations Are Not Interchangeable, But Are Complementary

A table allows for a powerful summary of data and makes comparison easier. But it also creates a need for a kind of interpretation best handled in prose sentences. The prose explanation is needed to point out the significance of the most important elements in a table.

The kind of emphasis available in prose is flattened in the table's own presentation of data. Every data point looks the same, so that none receives more emphasis than another. A good prose explanation thus complements the table, *completes* it, or rather completes the job of communication that the table can only begin.

From this *principle of complementarity,* we can derive *a principle of compensation.* It says that, *in every revision of a text, something is lost and something is gained.* The ideal text is an horizon, toward which we may move, but which we can never reach. In converting a paragraph into a table, a writer may gain accessibility and structural clarity, but may lose any kind of subordination and interpretive syntax not expressed in the x-y axis of the table.

Here, for example, is a prose account of a fictional weight watchers' group:

Bill weighed 200 pounds in January, 180 in February, and 170 in March. *Like Bill, Joe and Harry lost 20 pounds in the second month and 10 in the third month of the program.* Joe weighed in at 190 in January, dropped to 170 in February, and finished at 160. *In similar fashion,* Harry fell from 180 to 160 to 150.

Table 1 also gives an account of the program. Notice that, while the table gains considerable efficiency in reporting the weight loss, it flattens the emphasis by omitting the simple interpretive links provided in the italicized portion of the paragraph. To compensate for the loss, much of the italicized prose would have to be restored in a full report on the group. The table cannot totally replace the paragraph, but can only do part of its work. It can display data more efficiently than prose, but it can make no arguments, not even basic comparative arguments (such as "Bill is like Joe").

Table 1. Weight Record (by pounds) in
Three-Month Program (example)

	Jan.	Feb.	March
Bill	200	180	170
Joe	190	170	160
Harry	180	160	150

THE SEMIOTIC MODEL

As an extension of the principles of complementarity and compensation, semiotics offers a unified approach to the selection and composition of visual and verbal elements in technical documents. As we observed in Chapter 2, C. S. Peirce not only developed a theory of signs in general, but also came up with a classification of signs.

Peirce defined three general kinds of signs—the icon, the index, and the symbol. We can use this classification to arrive at a fuller understanding of how one sign relates to another and how different kinds of signs, in combination, create complementary and compensatory effects in technical documents.

We begin with the question, what are the kinds of effects created by different signs?

What is Representation?

First, to review briefly, every sign is an irreducible relationship of three elements—the object, the representamen, and the interpretant. This triadic relationship is reviewed diagrammatically in Figure 11.

Against the background of Chapter 2, the elements of the sign may be defined thus:

- The *object* is a thing, an action, a concept, or another sign that is represented in the sign.
- The *interpretant* is the effect upon a state of mind, text, or act—in the terms of pragmatism, the *meaning*—that arises during semiosis.
- The *representamen* is the medium of representation and interpretation, that which *substitutes* for the object and *determines* the interpretant.

The sign relation simultaneously permits three "grammars":

1. The representamen represents the object to the interpretant.
2. The interpretant interprets the object by means of the representamen.
3. The object determines the interpretant by means of the representamen.

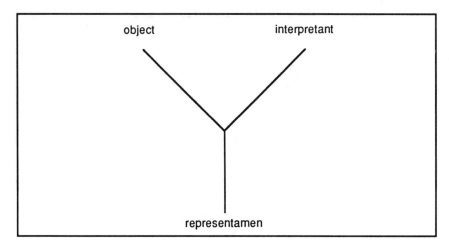

Figure 11. Peirce's triadic sign.

Peirce's formulation of the sign as an irreducible triad suggests that all of these actions are present at once. But to distinguish among icon, index, and symbol, we need to focus on grammar number 1: *The representamen represents the object to the interpretant.* And we need to concentrate especially on the first part of this sentence, *The representamen represents the object.*

Peirce defined *represent* as "To stand for, that is, to be in such a relation that for certain purposes [one thing] is treated by some mind as if it were [another thing]" [8 vol. 2, p. 273]. By asking the question, "In what way does the representamen stand for the object?," we discover how to distinguish among the three types of signs.

The Icon

The icon is, according to Peirce, a "diagrammatic sign . . . which exhibits a similarity or analogy" to its object [8, vol. 1, p. 369]. Thus, a portrait *resembles* a person, or a flow chart works by an *analogy* of a chronological sequence of events understood in terms of the natural motion of a liquid. Figure 12 gives a diagram of iconic semiosis, the process by which icons signify their objects.

The term *icon* is widely used in technical writing to indicate a small picture that substitutes for a number of words. For example, a picture of a folder substitutes for the word "file" in a computer's user interface. The icon thus makes explicit the analogy between a collection of data (letters, memos, etc.) in an office environment and a collection of data (electronic signals) in computer memory. This analogy suggests not only an image but also a use for the stored information. Although the information is invisible and intangible, users are able to employ it

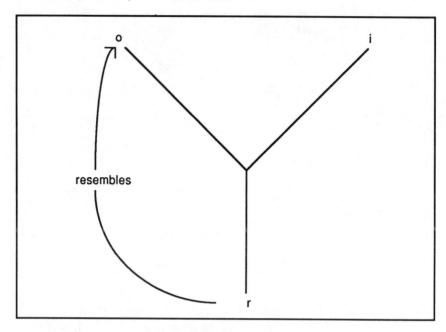

Figure 12. Iconic semiosis.

much as they would use the information stored in a company's office files. The icon is thus user-oriented.

Note that the term *file* is already an implied analogy—a verbal analogy, or metaphor—before the icon is presented pictorially. Words that are already highly suggestive of visual imagery can be transformed into icons with relative ease. The advantage of having the icon is that readers might save one step in translating the verbal image into a pictorial one in their own minds or may gain a surer understanding of the analogy with the aid of the icon. The writer who uses the icon thus does more of the work of interpretation for the reader. Without the icon, the reader may in fact overlook the analogy implicit in the word *file* and may simply think it is a technical term in computer language that merely bears a resemblance to an ordinary word.

The Index

The index "forces the attention to the particular object intended without describing it" [8 vol. 1, p. 369]. Like a pointing finger (the *index* finger), an index points to its object. Moreover, it frequently has a dynamic or existential relation to the object, such as proximity [8, vol. 2, p. 284, 299]. This means that the index is often physically close to its object. It draws attention to the object without re-determining its meaning.

The index is thus a purely rhetorical way of showing the reader what is important or a means of helping readers find their way around in a text. A pointing finger indicates something close to it that is worthy of attention. A pronoun locates its referent ("he" refers back to "Mr. Smith"). A symptom is a sign for the presence of a disease (a cough may come from a diseased lung). Figure 13 is a diagram of the relationship.

"Information locators" of various kinds are the primary indexes in a technical document. The table of contents, running heads, chapter numbers, and of course the index itself are all devices that orient the reader to the subject matter and provide a map of the document's territory.

Headings—such as "Results," "Discussion," and "Conclusion"—orient the reader structurally. However, as headings become *headlines,* they serve not only a structural but also a descriptive function, such as "A New Programming Language Applied to the Problem" or "How to Format Using the Dial Function." Because they describe their object as well as locating it, such headlines go beyond their indexical function and begin to work symbolically.

The Symbol

A symbol "signifies its object by means of an association of ideas or habitual connection between the name and the character signified" [8, vol. 1, p. 369]. Thus,

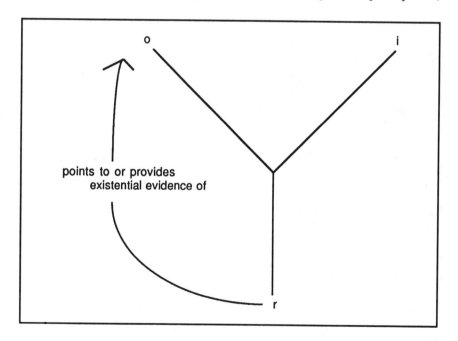

Figure 13. Indexical semiosis.

a noun denotes a person, place, or thing even though there is no necessary connection between the word and the object signified but only a conventional connection. If a speaker says "tree," an English-speaking listener will think of the green leafy thing, but a Pacific islander will draw a blank, unless the speaker adds an index (by pointing to a nearby palm) or an icon (a picture of a tree) to interpret the symbol. Figure 14 gives a diagram of the symbol's characteristic action.

For now, we can think of the written text as primarily symbolic, since language is the symbolic medium *par excellence*. The connection between the word *dog* and the four-legged creature that sleeps on the carpet by the fire is a purely conventional connection that must be learned and that will vary from culture to culture.

It is important to remember that icons and indexes help in the learning process. In a child's book, the word *dog* is placed beside a picture of the furry creature—an icon. And one of the first questions that children learn to speak is some version of "What's that?" or "What's this?" The child usually combines the demonstrative (indexical) pronouns "this" and "that" with the pointing index finger.

Theoretical Difficulties

Unfortunately, two major theoretical difficulties hamper our simple scheme and force us to pursue our theorizing a little further.

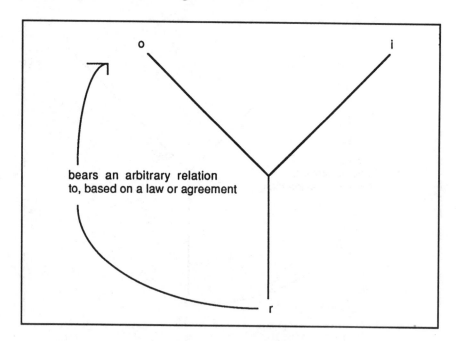

bears an arbitrary relation
to, based on a law or agreement

Figure 14. Symbolic semiosis.

1. *All signs in a document are to some degree conventional and there-fore symbolic.* From the perspective of document design, one problem with the Peircean scheme, a problem that Peirce himself recognized, is that all signs appearing in a document are technically symbolic, since they all depend upon conventions of discourse.

Line drawings, for example, are certainly iconic, but they resemble their objects in ways that are highly conventional and that must be learned. A child must learn a rudimentary science of visual relations before recognizing that the drawing in Figure 15 represents a rabbit, for example. Viewers can only grasp the relations of the drawing's highly simplified elements—the placement of the long ears and round tail, the relative size of the body's divisions—if their minds and eyes have been trained to interpret those relations.

The philosopher Susanne Langer points out that, unlike human children, dogs do not seem capable of interpreting pictures. They fail to recognize portraits, even photographs, of their masters. This is not because, as is commonly believed, they can't smell the images. After all, they will give their eager attention to the activities of a cat viewed through a window. Rather, they fail to recognize the portraits because they have not learned, and cannot learn, to relate two-dimensional representations to the world of three dimensions in which they live [9, p. 72]. And, unlike children, animals rarely show interest in television, again because they do not have the conventions of interpretation available to the human viewer.

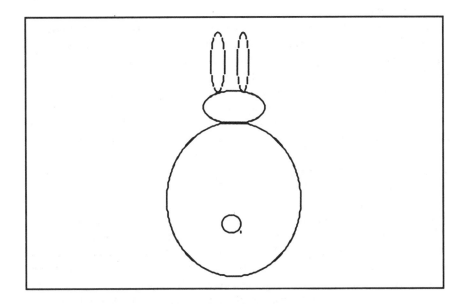

Figure 15. Bunny (example).

To maintain technical accuracy and still benefit from the potentially useful distinction of icon, index, and symbol, we may, following a suggestion by Peirce, designate the categories of conventional signs as *iconic, indexical, and symbolic,* understanding that they are all symbols but that each preserves a vestige of the ideal sign typology. Thus a sign is:

- iconic when it represents its object "mainly by its similarity" to that object [8, vol. 2, p. 157],
- indexical when its primary function is to locate other signs or objects,
- symbolic when its main function is to give meaning to its object.

2. The definition of icons as resemblances or analogies is inadequate. The simple solution we've developed to counter the first difficulty is not sufficiently rigorous, however, because the overlap in categories is more radical still. The category of the iconic is particularly troublesome. Consider two examples of icons that Peirce comments upon.

First, the diagram: Peirce admits that "many diagrams resemble their objects not at all" [8, vol. 2, p. 159]. For instance, the collections of lines and arrows we use to represent signs and semiosis in Figures 12, 13, and 14 have no natural resemblance to what they signify. But, Peirce maintains, "It is only in respect to the relations of their parts that their likeness consists." We are trying to give a shape to an abstract concept, but the shape does not exist in nature (except perhaps as a pattern of brain waves).

Thus, despite Peirce's objections, we must admit that the icon cannot strictly be said to resemble or bear an analogy to what is not there. It is a new image based on conventions of proportion and arrangement and is in this sense purely a symbol— the possible exception being the pointing arrow-like lines in Figures 12, 13, and 14, and indeed these arrows function more as indexes than as icons.

Our second example is the photograph. Peirce writes: "Photographs, especially instantaneous photographs, are very instructive, because we all know that they are in certain respects exactly like the objects they represent. But this resemblance is due to the photographs having been produced under such circumstances that they were physically forced to correspond point by point to nature. In that aspect, then they belong to the second class of signs, those by physical connection [indexes]" [8, vol. 2, p. 159].

Photographs, regardless of their strong resemblance to the objects they represent, are technically indexes. Indexes have an existential connection to their objects; thus, a pointing finger indicates something close by; a barometric reading may indicate an approaching storm; spots on the face and body may indicate measles. And the image on a photographic plate indicates an original proximity to the object represented. In this sense, the photographic image is not an icon, despite its near perfect resemblance to its object.

Hence, the very definition of the icon as a sign that resembles its object is placed in doubt. The sign that seems to be most appropriate to the category, the photograph, is in fact an index, not an icon. And signs that seem more appropriately placed in other categories because of their conventionality, such as theoretical diagrams, are in fact good examples of icons [10].[1]

Solution: Substitute Sign-Function for Sign-Unit

Have we now taken back everything we have given? Not if we make one final adjustment to our model. We must simply remember that signs are not technically *units,* but are rather *functions* within a text.[2]

An individual textual element may, depending on the context, function in a variety of ways. Its function may be primarily iconic, indexical, or symbolic, depending upon its relation to the other signs in the text and in the context of action preceding or resulting from the text.[3] A picture may tend to be iconic, but in certain circumstances it may function as an index or a symbol. The attribution of sign-function to a textual feature is part of the process of meaning-making, interpretation.

Peirce himself recognized that the nature of a sign is determined by its *function* within the *process* of semiosis. We have, therefore, done an injustice to Peirce's theory—although it is an injustice for which he himself provided a precedent—by treating the relation of object and representamen in isolation from the relationship with the interpretant, the effect upon a mind or principle: The representamen represents an object *to an interpretant.* It is the interpretant that determines the result of the interaction of object and representamen. The result of the interaction is the total sign-function.

What are the practical effects of this re-reading of Peirce? For one thing, it keeps us from oversimplifying the distinctions among icons, indexes, and symbols by categorizing words, photographs, diagrams, tables, and pointing fingers into an all too neat set of instructions about how each textual element will perform a discrete communicative task. We must instead lay down a set of functional relations, which at various times may be assigned to different textual elements.[4]

By revising Peirce's notion of iconic signs, we can render a simple workable theory that will serve as a makeshift guide for understanding the relationship of the various elements and functions of a document. In general, we can now revise the descriptions of the three kinds of signs in our earlier diagrams, by adding in Figures 16, 17, and 18 a functional definition of each relation (given in parentheses after the original description).

Our general theoretical conclusions, then, may be summed up as follows:

- Iconic signs are *relational and holistic.*
- Indexical signs are *locational.*
- Symbolic signs are *analytical and interpretive.*

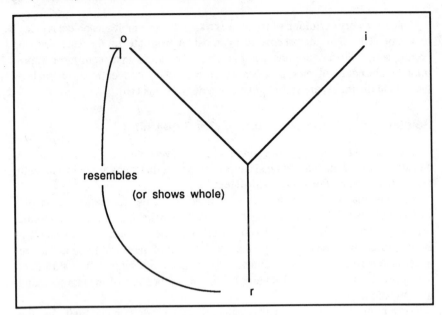

Figure 16. Sign-function in iconic semiosis.

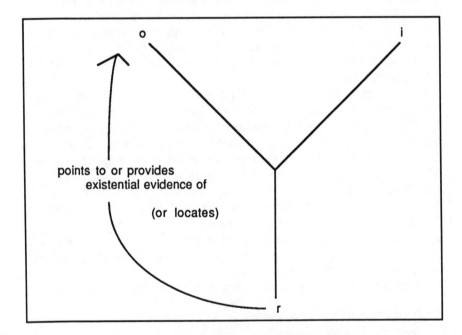

Figure 17. Sign-function in indexical semiosis.

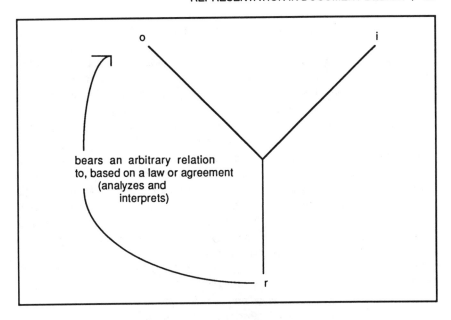

Figure 18. Sign-function in symbolic semiosis.

RESTORING THE INTERPRETANT:
THE COGNITIVE APPROACH

We are now in a position to apply the model to an analysis of some realistic communication situations involving total document design. To do so, we need only to develop contexts.

Contexts are the environments in which signs or texts function. The text itself is a context for the individual signs within it. The context for the text as a whole may be *psychological,* the mental field in which representation and interpretation takes place (the mind of a writer or reader); or the context may be *socio-cultural,* the field of human action that surrounds and supports the text. The beauty of the semiotic model is that it helps us to describe, analyze, and manage the relation of one kind of context to another (for they are, as we will show, deeply interrelated semiotically).

We will begin with the cognitive approach, which emphasizes psychological contexts, considering how each sign-function potentially makes an impression upon the human mind and how one function complements and compensates for the others in this mental action.

Iconic Sign-Functions

Iconic signs present general impressions that allow the observer to view a thing or a concept as a whole. In technical communication, most diagrams, tables, and charts function iconically (though they may have an indexical function as well).

Graphic theorists who draw on cognitive psychology, like Rudolf Arnheim, often say that the value of an image is that it allows a viewer to attain an impression of a whole that would be broken into pieces if it were presented in the relatively linear and atomistic medium of language [11, p. 249]. This is especially true in the case of line drawings and photographs, which give a viewer a quick impression of the whole surface or outline of an object—a tree, for example, which written language must either classify ("maple") or break into component parts and attributes ("a large hard wood with broad leaves and rough bark").

Images are primarily synthetic; language is primarily interpretive and analytical. A written text depends upon an analysis of various bits of the whole into discrete symbolic units—words. The same is true of mathematical texts. But images make the whole present at once to the viewer. From a cognitive perspective, this relational power is the great advantage of iconic signs.

Icons, then, have a power to synthesize that is absent in more purely symbolic media. To take another example, we might argue that our diagram of the semiotic triad (reproduced in Figure 19) is the only way to represent the sign relationship as irreducible. Obviously, we can't absorb the icon all at once, but have to read "object," "representamen," and "interpretant" separately. The image is thus a hybrid of icons and symbols.

But we are better able to grasp the simultaneity of the relationships than if they were expressed only in sentences (or equations). If we say, "The representamen represents the object to the interpretant," we must speak of each element and each relationship one at a time. We must break the sign into parts so that it will fit into

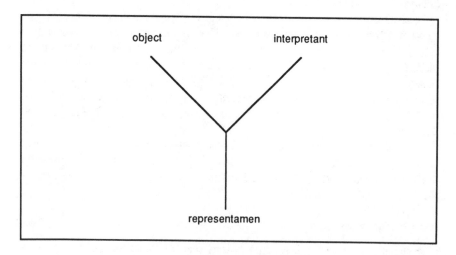

Figure 19. Peirce's triadic sign (reproduced).

a sentence, where the reader will encounter in linear fashion first the subject ("representamen"), then the verb ("represents"), and in turn the direct object ("object") and the object of the preposition "to" ("interpretant"). One element inevitably precedes the others (in this case the subject "representamen"), thereby gaining an artificial (or unintended) ascendency or prominence. In the diagram, however, the parts of the sign are arranged in a way that suggests their relative equivalence.

Likewise, the organization chart in Figure 20—a graphic outline of menu options from a hypothetical software manual—gives an overview of the system that could not be realized in the language of the manual and that also is unavailable online since the three separate menus could not be viewed simultaneously on the computer screen. The iconic power of the chart depends upon its ability to do what other media cannot—show the relationships implicit in its object as a whole. This diagram allows the reader to view all the menus at once, which would be impossible not only in language but also in the computer interface itself. The diagram thus complements both the prose of the manual and the user's experience in the interactive medium provided by the computer.

In addition to their practical usefulness, icons may play a major role in the making of theoretical models. Cognitive psychology suggests that people use iconic signs to hold theoretical conceptions in memory. At best, we can retain only a few sentences in the mind at one time. But we can retain and even manipulate the elements of a single diagram—or an outline, which is in fact a kind of a hierarchical diagram, as Peirce himself suggests [8, vol. 2, p. 282]. In several places,

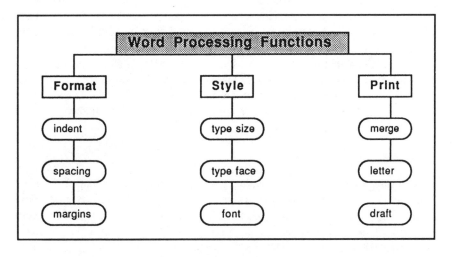

Figure 20. Diagram showing multiple menus of word processing functions (example).

Peirce claims that most high-level thinking is originally iconic or diagrammatic, a point affirmed in Arnheim's treatment of scientific modelling [11, p. 116].

The great strength of iconic signs is their power to impart a total impression of objects and to show relationships of parts to the whole. Their weaknesses are connected with their inability to locate their objects in an overall context (which is the function of indexical signs) and to interpret and analyze the objects they present (which is the function of symbolic signs). That is why illustrators must often use "call-outs"—lines that connect words to the picture—to point out important pieces of the whole image.

Indexical Sign-Functions

Call-outs are indexical signs; that is, they allow the reader to locate and contextualize information. Other indexical devices include tables of contents, headings, and running heads. Typographical variations—shifts in typeface, type size, and font—function similarly and also enhance emphasis.

Indexical signs reveal their greatest value in their use as devices for providing emphasis. They give the writer a way of saying, "Here's something different or important."

But, in addition to guiding the reader *textually,* indexical signs also function *contextually* [12].[5] Indexes situate the reader in relation to the world outside of the document and are therefore crucially important in an action-oriented discourse. This kind of contextualization is the primary function of orientation photographs, maps, and flow charts.

Perhaps the best example in technical communication is the simplified line drawing in a user's manual that highlights a special area being discussed, as in Figure 21. This diagram directs the reader's attention to darkened spaces on a map of the machine, which corresponds to the location of special keys under discussion in the written text.

In a statement that recalls Peirce's indexical pointing finger, Arnheim speculates that line drawings may have originated in physical gestures. Indexical signs point to important places in the text and in the world outside the document, to which the reader must eventually turn to perform the actions inevitably proposed by technical documents.

Symbolic Sign-Functions

According to Peirce, icons and indexes "assert nothing" [8, vol. 2, p. 291]. They show, but they do not tell. Symbolic signs, however, are telling signs; they allow for analysis, interpretation, and judgment. They constitute the bulk of the written text. Although words may also function indexically (in headings and cross references) and even iconically (in figures of speech—like the metaphor *file*), language in the symbolic mode is the primary vehicle for analyzing information, breaking it into component parts and assigning relative values to each part.

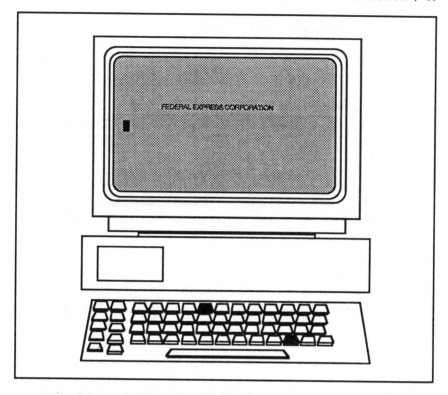

Figure 21. Indexical diagram of computer with important keys
highlighted in black (example).

One important side effect of this function is a reduced pace. One study of "wordless instructions," for example, concludes that reader comprehension may be inhibited by the fast pace of instructions composed entirely of pictures and diagrams [13]. Words can compensate for this tendency of icons to overload the reader's perception. As there are benefits to holistic presentation, so are there benefits to analytical presentation. Hence the principle of complementarity.

Symbolic signs also have the power to focus attention in a way not available with the use of other kinds of signs. When verbal call-outs are added to photographs and diagrams, the writer creates textual elements that not only act as descriptors but also draw attention to important details and force attention away from unimportant ones, coupling indexical power with the power of identification and judgment. Likewise the written text that accompanies a table, chart, or any other high-density iconic sign should perform the crucial task of guiding the reader's judgment by focusing attention on the most important data points in a careful analysis that anticipates conclusions.

Analyzing the Effects of Textual Revisions

An awareness of the types of cognitive effects produced by the three types of sign-functions enables us to ask the most important question of document design, a question motivated by the principle of compensation: "What will be gained and what will be lost if I make a certain change?"

The shift from plain prose texts to iconic-mosaic designs comprises a shift from symbol-heavy writing to a more semiotically varied form of writing. What are the consequences of such a shift? What is lost and what is gained?

Let's take, as an example, a very common revision in modern technical writing—the use of numbered or bulleted lists in the place of a regular prose paragraph. A list increases the accessibility of information and makes it more readable. Such is the intention of the following list of risk factors from an article about cancer prevention in a government publication:

- alcohol
- tobacco
- diet
- radiation
- workplace
- estrogen

The list is used to structure the entire article (which space does not permit us to reproduce as a whole). In the bulleted format we give here, the device appears first in the introduction. Then each element of the list is used as a heading of a major section.

The rhetorical integrity of the list demands that each entry be reduced to a single-word noun. Lists must be grammatically parallel. For the browsing reader, some of the information will be adequately conveyed by this extreme grammatical limitation; "alcohol" and "tobacco" indicate almost immediately the actions that the reader must avoid to stay clear of the risk of cancer. But what about the others? "Diet," for example, would be more adequately expressed by "Eating Foods High in Fat and Cholesterol." "Radiation" would be more adequately expressed by "Spending Too Much Time in the Sun."

This is not to say that the list is a failure, but only that, because of the limitations placed by the grammatical austerity of the list, the writer must use other rhetorical techniques to compensate for the list's inherent weakness. The first sentence in the sections headed "Diet" and "Radiation"—the most emphatic sentence position—contains the clarifying information. The article also includes cartoon images (icons) beside each element in the list to hint quickly at the full meaning of each term. The icon for "radiation," for example, shows people lying on beach blankets in the sun.

Consider one more example: the replacement of sentences with high density graphics like tables and charts. Such graphics contain so much information that they threaten to overwhelm the reader or to present information at too fast of a pace. We demonstrated earlier in this chapter how the prose that accompanies a table can be crucial in providing a guide for how to read and interpret the information provided. The prose text provides a model of reading behavior. By providing this guiding prose, the writer compensates for possible reader saturation and the possible loss of the argumentative and narrative contexts. The symbolically and indexically functioning prose complements the iconic power of the table.

In sum, then, restoring the interpretant, from the cognitive perspective, is a matter of making sure that there are enough links in the chain between the reader and the original object of the text. This rationale lies behind the action of a writer who, in trying to describe a machine, presents an iconic photograph, an indexical arrow, and a symbolic callout, and maybe even a brief description in the written text as well.[6] The iconic-mosaic combination of these signs allows the reader to get oriented to the machine in such a way as to act in the prescribed manner.

Restoring the interpretant means making sure that the communication media establish the proper contexts for information so that readers can find their way around in a text and can see the relation of the text to the external world.

RESTORING THE INTERPRETANT:
THE CULTURAL APPROACH

A possible shortcoming of the cognitive approach is that, perhaps naively, it assumes a particular kind of audience—a reader interested in efficient learning or gathering of information—a busy, hard-working active reader, the ubiquitous problem-solver that figures so largely in American business, engineering, and science—a reader who, in semiotic terms, becomes an interpreter devoted to converting writing to further technological action. For a large percentage of rhetorical situations in the field, these assumptions about the kind of audience addressed are quite adequate to the task of producing effective documents. But can there be any other kind of reader in technical communication, and how would such a reader be effectively addressed?

Culturally oriented studies suggest the possibility of a more pliant and variable interpreter, a reader who is in many ways more passive, more open to manipulation than the active reader posited by cognitive studies. Attempts to address this reader in technical documentation, or even to create or encourage such readership, have not been the subject of careful analysis in theoretical writings about technical communication and the rhetoric of technology [14].[7] A cultural perspective, which recognizes the socio-political context of technical writing and builds that context into the semiotic model, helps to explain recent trends in addressing alternative readers of such documents as corporate annual reports, technical sales literature, and user manuals.

Brochuremanship and Reader Manipulation

Partly as a result of the close interaction of technical communicators with marketing personnel and media consultants, many design techniques in recent general-market technical documents have been borrowed from the mass media, in particular from advertising.

We have said, following Marshall McLuhan, that the chosen layout of many modern technical documents is iconic and mosaic. According to McLuhan's view, fragmented images invite the reader of advertisements to gather an impression without analyzing or evaluating the information presented; iconic signs prevail over symbolic signs because the message is not intended to be thoughtfully interpreted [15-17].[8] Viewers are offered instead a "subliminal pill" [7, p. 229], whose effect will be felt at the time these consumers make a decision about what to buy.

In certain kinds of technical documents, such representational trends are, of course, wildly inappropriate and are sometimes actively discouraged. In specifications for technical proposals, for example, there has been an active (but only partially successful) attempt to curb the trend toward "brochuremanship."

The proposal, like advertising, is a direct attempt to influence business decisions. But, though a technical proposal may also appear "mosaic" in its mixing of icons, indexes, and symbols, and though it may begin as a graphically oriented storyboard [18, 19], a compositional technique pioneered by advertising specialists, the proposal can never be primarily iconic without losing focus and relapsing into incoherence. Proposal writers rely much more heavily on formal argument than do ad writers. Moreover, brochuremanship is expressly forbidden in many types of proposals to the government. Regulations against the use of "slick" graphics and formatting represent official efforts to keep such techniques from unfairly biasing evaluators of the proposed projects [20].

In the nonregulated marketplace, however, technical writers have drawn widely upon the manipulative techniques of the mass media. Many teachers and practitioners are likely to look on this evolution with suspicion: Isn't the goal of technical writing precision and clarity, and doesn't the incorporation of the techniques of the ad writer undermine the integrity of a document?

We would argue that it is best, at least from a theoretical perspective, to defer such doubts. After all, brochuremanship exists; and, whether or not we choose to practice its techniques ourselves, we need to describe them and explain their significance in the field rather than merely censuring them from the start. Furthermore, instead of applying preconceived ideas about what constitutes clarity and precision, a pragmatic semiotics encourages us to question the general effectiveness of brochuremanship within specific contexts. If it works there, can we realistically question its "clarity"?[9]

An Example: The Sleek Annual Report

For example, many major companies now spend great amounts of money to develop impressively printed annual reports. Icons prevail. If the company desires to create a strong regional identification—perhaps most of the stockholders are from the local city—then photographs of local landmarks are juxtaposed with company icons. A report from a large Memphis company, for example, could very well show a delivery truck sporting the company logo and driving along Riverside Drive, the Mississippi River in the foreground, the skyline of "new Memphis" in the background. The icon would speak of tradition *and* innovation—the river *and* the skyscrapers—in an effort to appeal to progressives and conservatives among the stockholders.

In fact, however, in the 1988 annual report of one big Memphis-based company, the Federal Express Corporation, the designers opted for an international appeal, deciding to represent the company's growth and the spreading of its influence instead of its pride in its local identification. Not coincidentally, a merger with a large west-coast shipping concern was announced soon after the report appeared.

The cover featured a world map done up in brown wrapping paper stamped with the Federal Express seal. The producers of the report spared no expense. They used four colors in the printing, not only to provide for easy reading and high impact, but also to suggest prosperity: The company can afford expensive printing—only the best for its loyal stockholders.

The language of the report is likewise the best classic business writing that money can buy. Here is the first paragraph in the section of the introduction entitled "To Our Stockholders" [21, p. 3]:

> With revenues nearing $4 billion, Federal Express remains the leading company in the express delivery industry. During 1988, our employees around the world handled over 226 million packages and documents—a nightly average of approximately 900,000 and a substantial increase over the previous year.

The writing is certainly good—note especially how the simple phrase "around the world" picks up the theme of the cover illustration—but the style is plain, conventional, direct business history.

The writers can afford a style that is no frills and all business. They do not have to overstate their case or stretch the truth, because the sensationalistic effects of the document are brought off by the graphics. Bar charts fulfill the reader's expectation that the report be "technical" and "professional." But they are by no means ordinary bar charts, which would seem dry and dull; they are *four-color bar charts* that carry the connotation of richness and flare.

The graphics do most of the rhetorical work of the report, and for good reason. What will most readers *do* with the report? Will they really read it or simply look it over?

Consider the remarks of graphic designer Jan V. White: "What gratifying illusions of grandeur blind you when you unwrap that sleek annual report? How startling is this year's typeface on such impressive stock? Marvel at the lightness and precision of the strokes. That bespeaks quality. This is your company (never mind that you hold only forty shares). It feels good to have arrived. You'll keep it right there, on your shelf of honor" [5, p. 7].

Connotative Value of Signs

Whereas the cognitive approach to the interpretation of document design emphasizes the *denotative* or literal value of sign functions, the cultural approach focuses on the *connotative* value—the *feel* of the thing. The best work on the semiotics of connotation has been done by the French theorist, Roland Barthes, first in his ground-breaking *Mythologies* [22] and later in his *Elements of Semiology* [23].[10]

Denotations, Barthes argues, are relatively continuous and stable meanings of signs, but connotations are "discontinuous and scattered"; they are "fragments of ideology," politically loaded elements of a "second-order signifying system" that hover over the denotative meaning of a term [24, pp. 89-92].

Whereas denotations tend to be stable across a wide range of readers, the interpretation of connotations depends upon varying perspectives and contexts. The noun *American,* for example, denotes quite simply a person from America. But the simple denotation masks a source of deep emotion whose effects vary with different contexts and audiences. Imagine how the word would be received at an American Legion rally in Peoria, on the one hand, or at a political rally in Iran, on the other. In one context, the word would connote pride and tradition; in the other, it would connote oppressive power that must be resisted.

If such political examples seem far removed from the world of technical writing, consider how carefully contemporary user manuals for microcomputers must manage terms and concepts like *power* and *control.* When the user is made to feel powerful and in control, such terms have very positive connotations; but if the power is said to reside in the machine or in some abstract system to which the machine connects, the user is likely to feel threatened, lost, or inadequate.

The "user-friendliness" movement in computer documentation has tried to create an image of a docile machine ready to perform any task for its master, the user. This marketing rhetoric represents a technical communicator's rewriting of the old salesperson's cliché: "The customer is always right" becomes "The user is always in control."

Icons are particularly liable to variations in connotation because their denotations are less likely to be stable than are the denotations of words, which are established by grammar and usage and which can be decoded with the help of dictionaries. Barthes asserts that, unlike words, whose range of meaning is limited by rules, images invite varying interpretations.

If a picture is to have a very specific meaning, the writer has to find a way to connect the iconic and symbolic functions of a text [24].[11] Every culture develops conventions to meet this discourse need. Thus, cartoons will contain bubbles which allow the characters to "speak," a photograph will be accompanied by a caption, and a line drawing will have arrows leading to verbal call-outs and even an accompanying discussion in the written text.

Some images have been so consistently used by the mass media, however, that their meanings have stabilized for the historical moment. The cowboy on horse-back in the Marlboro cigarette ad bespeaks macho individualism, the kind that encourages risk-taking, that all but laughs at the surgeon general's warning about the health hazards of smoking. Part of the work of advertising is to control the fixing of connotational meaning in such images. The ad writer's hope is that a culture immersed in a flood of images will respond in relatively predictable ways.

An Example: A Connotative Users Manual

In technical writing, an interesting case in point is found in Hayes Corporation's award-winning introductory volume to the *Please* program. Usually such volumes are titled "Before You Use Your Program" or "Getting Started." For this piece, Hayes landed on the concept of "Ready, Set, Go!" which draws on the American institution of athletics as a metaphor for all forms of competition and achievement.

The cover of the little manual sets the theme by showing a sprinter rushing to get out of the blocks at the start of a race, much like the ideal user who cannot wait to begin working with the program. The connotations of impatient competitive-ness and willing strength predominate and sharply contrast with the polite title of the software itself—*Please.*

The style of the illustration (a print, probably produced in the original drawing by colored magic markers) suggests the painting style of post-impressionism, particularly the work of Cezanne and Van Gogh, with its restless creativity—an inviting source of identification.

The facing page of each chapter manipulates the theme by depicting the runner in various attitudes. The section on preparation, for example, finds the runner doing preparatory stretching exercises in a graphic appeal to the business user's implied understanding of and obsession with exercise. What lunchtime jogger would fail to do the proper stretches before beginning the workout? One area of expertise (business computing) relates to another (exercise technology) in true iconic fashion.

The effectiveness of such graphics is dependent upon the kind of cultural knowledge that marketing specialists long for. What will appeal to the fantasies and appetites of the audience? (McLuhan has wryly suggested that "advertising is the science of man embracing woman.") Will cartoons in manuals "soften" the threat of technicality and make the job of learning computer technology more appealing, or will they insult the adult intelligence of the user? Will banks of high-density data tables in annual reports suggest authority and computational strength, or will they merely intimidate the reader or connote a needless technicality?

Because the answers to these questions are so uncertain, there is a great deal of risk involved in the application of primarily iconic presentation in technical communication. But the trend of the recent literature indicates that many companies think it is a risk worth taking.

Mass Media and Technical Communication: A Case of Semantic Interpenetration

The passing of techniques from the mass media into technical communication is matched by the passing of technical rhetoric into the mass media: Consider the elaborate weather map and the quasi-technical charts of *USA Today*. This interaction of the two fields is a relatively recent phenomenon, whose rhetorical and cultural effects will only be determined by extended historical and analytical research. The interpenetration of these diverse forms of discourse suggests that effective interpretation of signs requires a reader not only to have the cognitive competence to reconstruct the pieces of the semiotic chain between objective and outcome, but also to possess cultural knowledge, a background of insights into the politics and social ideologies that form the context of the communication.

Technical writers have always worked (sometimes unconsciously) to accommodate their readers' most demanding historical needs. In the market of a couple of decades ago, there was little need for a self-sufficient, "user-friendly" computer manual since only highly trained technical staff used computers. In the early years of computer use in the office setting, the parent company often sent out a team of trainers to help everyone in consumer companies learn to use the machines. But, by the time the computer was mass-marketed, the manual that became part of the product package began to reflect the mass culture in which it was intended to operate. And, in turn, as the mass culture has become ever more "computer literate," the mass media has evolved to meet the expectations of an audience with increasing technological sophistication.

In addition, however, it is more and more common to see technical documents do more than merely accommodate historical trends. The tendency is also to manipulate the reader by fixing connotations in images associated more with Madison Avenue than with Silicon Valley.

In the 1970s, the influx of such images in proposals that market products to the federal government resulted in strict regulations against "brochuremanship" in contract proposals. In recent years, with the advent of "desktop publishing," the enforcement of such regulations appears to have slipped somewhat.

The lesson is that the very concept of "slickness" and "brochuremanship" is historically conditioned. The connotations of such words, and even their denotations, shift with time, and companies rush to control the refixing of their meanings.

From Signs to Codes

A classification of signs as textual elements—suggested in the differences of icons, indexes, and symbols—can help to determine how writers and their readers make meaning and can thereby serve as a means of mapping psychological and cultural changes in textual preferences. As signs combine in texts and fall under the influence of cognitive and cultural effects, they begin to form into codes, systems of communication that require special experience in reading and writing or membership in exclusive communities of knowledge.

In Part II, we will see how the concept of semiotic codes is useful as we turn to classifying types of documents—the genres of technical communication. The creation of generic categories for technical documents requires a further theorizing and specifying of various contexts and situations for communication.

NOTES

[1] The contemporary Italian semioticist Umberto Eco therefore rightly encourages us to "face the problem of the so-called iconic signs": "even though there is something different between the word /dog/ and the image of a dog, this difference is not the trivial one between iconic and arbitrary (or 'symbolic') signs. It is rather a matter of a complex and continuously gradated array of different modes of producing signs and texts, every . . . sign-unit . . . being in turn the result of many of these modes of production" [10, p. 190]. Eco concludes that the icon "is not a single phenomenon, nor indeed a uniquely semiotic one" [10, p. 216]. Moreover, he claims that this failure of iconism points up "a radical fallacy in the project of drawing up a typology of signs" [10, p. 217] and even that "the notion of a sign is untenable when confused with those of significant elementary *units* and fixed *correlations*" [10, p. 216].

[2] In this approach, we follow Eco, though not as far perhaps as he would like. "Properly speaking," he writes, "there are not signs, but only sign-functions" [10, [p. 49]. We may still speak of signs as units, but only as "the provisional result of coding rules which establish *transitory* correlations of elements, each of these elements being entitled to enter . . . into another correlation and thus form a new sign" [10, p. 49]. "Therefore the classical notion of 'sign' *dissolves* itself into a highly complex network of changing relationships" [10, p. 49].

[3] Peirce anticipated this very adjustment, not only in his attempt to deal with the overlapping of the icon into other categories, but also in this sentence: "no Representamen actually functions as such until it actually determines an Interpretant" [8, vol. 2, p. 175]. This is a negative way of saying that a sign becomes a sign only when it functions as such in the presence of an interpreting mind. Thus a picture is no different from a stick to a dog. A pointing finger becomes a sign only when someone connects it with the thing to which it points. A word is just a set of marks on a page until someone reads it as signifying a meaning. But none of these things—a picture, a pointing finger, a word—always necessarily functions as a particular kind of sign (icon, index, or symbol). The interpreting mind

must attribute one of the sign functions to it, either a part of text production (writing) or text consumption (reading).

[4]Though Eco provides the bare bones of a new theory of semiotic relations that would replace the notion of signs with the notion of codes, we feel that what he gives is not sufficiently worked out to be applied to the task of document production.

[5]As the semioticist Kaja Silverman notes, "Indexical elements help to transform general assertions into specific statements, to locate a discourse in relation to time and space" [12, p. 21].

[6]The reader may well remark at this point that every technical writer already knows to do this. Again, the point is not that theory reveals particular points of practice that have not been guessed before, but rather that theory shows how one point can be connected to another in a principled way and can thereby give a rationale for practice—a rationale that the writer can use, for example, to explain and justify actions to colleagues, something more to say than "That's the way it's done."

[7]Mass media studies, on the other hand, have been virtually obsessed with this general type of reader. For a good review, see "A Rhetoric of Mass Communication" by Lynette Hunter [14].

[8]McLuhan's view, which builds upon the seminal work of Vance Packard, especially Packard's famous book *The Hidden Persuaders* [15], is opposed by a more conservative, information-oriented interpretation of advertising rhetoric. This view, which receives its classic treatment in David Ogilvy's *Confessions of an Advertising Man* [16] and *Ogilvy on Advertising* [17], is more closely aligned with the cognitive approach to technical communication.

[9]This does not mean that we should overlook the issue of ethics in audience manipulation. Teachers and writers, like most middle-class Americans, are suspicious of advertising, but ironically they still associate its techniques with "professional quality" publications. The key to ethical performance is to remain critical of the ends of technical communication as well as the means. Plainness, after all, could serve manipulative ends as well as brochuremanship. It can, for example, convey an impression of honesty and straight-forwardness in documents founded on factual distortions and misrepresentations.

[10]Barthes' work nicely complements Peirce's. Peirce's general theory of signs provides the basis for effective culturalist analysis, but Peirce himself never pursued socio-political implications very far, though he was very interested in the normative effects of discourse communities. Like Aristotle, however, and despite his protestations to the contrary, the bulk of his writing about "pure rhetoric" focused on psychological effects.

[11]Barthes writes, "images are polysemous; they imply . . . a 'floating chain' of signifiers, the reader able to choose some and ignore others." "Hence in every society various techniques are developed intended to *fix* the floating chain . . . in such a way as to counter the terror of uncertain signs; the linguistic message is one of these techniques" [24, p. 39].

REFERENCES

1. E. Tebeaux, Visual Language: The Development of Format and Page Design in the English Renaissance Technical Writing, *Journal of Business and Technical Writing, 5*, pp. 246-274, 1991.
2. B. F. Barton and M. S. Barton, Toward a Rhetoric of Visuals for the Computer Age, *Technical Writing Teacher, 12*, pp. 126-145, 1985.

3. J. Kalmach, Technical Writing Teachers and the Challenges of Desktop Publishing, *The Technical Writing Teacher, 15*, pp. 119-131, 1988.

4. P. M. Rubens, Desktop Publishing: Technology and the Technical Communicator, *Technical Communication, 35*, pp. 296-299, 1988.

5. J. V. White, *Graphic Design for the Electronic Age: The Manual for Traditional and Desktop Publishing*, Watson-Guptill, New York, 1988.

6. M. J. Killingsworth and B. G. Jones, Division of Labor or Integrated Teams: A Crux in the Management of Technical Communication? *Technical Communication, 36*, pp. 210-221, 1989.

7. M. McLuhan, *Understanding Media: The Extension of Man*, McGraw-Hill, New York, 1964.

8. C. S. Peirce, *The Collected Papers of Charles Sanders Peirce*, 5 vols., C. Hartshorne and P. Weiss (eds.), Harvard University Press, Cambridge, 1960.

9. S. Langer, *Philosophy in a New Key*, Harvard University Press, Cambridge, 1942.

10. U. Eco, *A Theory of Semiotics*, Indiana University Press, Bloomington, 1979.

11. R. Arnheim, *Visual Thinking*, University of California Press, Berkeley, 1969.

12. K. Silverman, *The Subject of Semiotics*, Oxford University Press, New York, 1983.

13. R. J. Scofield, Can Graphics Do the Entire Communication Job? *Proceedings of the 24th International Technical Communication Conference*, Society for Technical Communication, Washington, D.C., pp. 174-176, 1977.

14. L. Hunter, A Rhetoric of Mass Communication: Collective or Corporate Public Discourse, *Oral and Written Communication: Historical Approaches*, R. L. Enos (ed.), Sage Publications, Newbury Park, California, pp. 216-261, 1990.

15. V. Packard, *The Hidden Persuaders*, D. McKay, New York, 1957.

16. D. Ogilvy, *Confessions of an Advertising Man*, Atheneum, New York, 1964.

17. D. Ogilvy, *Ogilvy on Advertising*, Crown, New York, 1983.

18. R. Green, The Graphic-Oriented (GO) Proposal Primer, *Proceedings of the 32nd International Technical Communication Conference*, Society for Technical Communication, Houston, pp. VC 29-30, 1985.

19. M. Zimmerman and H. Marsh, Storyboarding an Industrial Proposal: A Case Study of Teaching and Producing Writing, *Worlds of Writing: Teaching and Learning in Discourse Communities of Work*, C. B. Matalene (ed.), Random House, pp. 203-221, 1989.

20. R. R. Oslund, Brochuremanship—How Much Is Too Much? *STWP Review, 12:3*, pp. 14-15, 1965.

21. Federal Express Corporation, Annual Report, Memphis, Tennessee, 1988.

22. R. Barthes, *Mythologies*, A. Lavers (trans.), Hill and Wang, New York, 1972.

23. R. Barthes, *Elements of Semiology*, A. Lavers and C. Smith (trans.), Hill and Wang, New York, 1968.

24. R. Barthes, *Image, Music, Text*, S. Heath (trans.), Hill and Wang, New York, 1977.

PART II
Genres

CHAPTER 4

Genres of Technical Communication

GENRE AS A MEANS OF REPRESENTATION AND INTERPRETATION

We have now shown that, as the rhetorical function of a text changes, the types of signs and combinations of signs that work best in a text will also change. In this chapter, we'll look at a variation on that theme: As rhetorical function and context shifts, so the overall type of document, the *genre,* also shifts. This shift has profound effects upon reading, writing, and technical action.

What Is a Genre?

The concept of genre is more than just a way of distinguishing among types of documents. It is a way of classifying and predicting types of action by determining the rhetoric of documents most closely associated with those actions. It is a coded way of suggesting appropriate roles for authors, audiences, and their mutual actions.

In selecting appropriate genres, writers model appropriate actions. They tell the reader something about the action implicit in the document and something about how they want the reader to respond.

The major genres of technical communication—the report, the manual, and the proposal—all approach technical action from a different perspective. The most obvious difference involves time. A report deals with *past* actions, actions that the reader may repeat or improve upon. A manual is about *present* actions, actions that may even accompany reading. A proposal projects actions into the *future,* actions that the reader must be persuaded to accept.

For each of these genres, writers must select a significantly different set of signs. In this chapter, and in Chapters 5 and 6, we will explore at length the differences between the major genres of technical communication.

What Are the Effects of Shifting Genres?

The following case suggests how shifts in genre affect reading and writing:

A team of soil specialists has done a study of local farmland and must present two reports—one in a refereed journal to be reviewed by their scientific peers and another for a group of farmers at a meeting of the local cooperative.

The presentations will differ immensely. In the journal article, the scientists will not have to explain the meaning of many of the technical terms and concepts that will need to be defined (or omitted) in their workshop for the farmers. In the article, however, they *will* have to provide a great deal of information about the land they are studying—its topography and general condition—information that scientific readers will not automatically possess. In the presentation to the farmers, on the other hand, the general information about the terrain can, of course, be omitted since it is part of the audience's background knowledge.

The scientists must therefore accommodate the presentation to the knowledge and needs of the audience. To do so, they must also understand the interpretive habits of the different audiences. The habits of peer reviewers, the first audience for the journal article, are influenced by the formal methodology and theory of "normal science" (see Chapter 2). The farmers' interpretations are more likely to be molded by experience and custom as well as by education in formal systems.

The nature of the signs appropriate for each audience will alter accordingly. In addition to the types of charts, graphs, pictures, and language that the scientists use, they will have to consider carefully the "voice," "persona," or "self" that they put forward. A very different authorial *image* will emerge in each of the presentations because of the audience's interpretive habits and the authors' efforts to accommodate what is generally understood about those habits.

For instance, the scientists may believe that farmers tend to be both curious about and suspicious of the knowledge of experts and may thus try to build identification by speaking of their own farms or of their background in farm life. This sort of autobiographical information would seem inappropriate, even silly, in the journal article. (Of course it may fall flat with the farmers, too, who could view it as so much empty rhetoric put forth by patronizing experts.)

The way the farmers will respond also depends upon how the information is presented. At the end of the scientists' talk, the farmers might well ask, "Are you proposing that we stop using chemical fertilizers altogether?" The experts might respond: "We're not proposing anything at all. We're just giving you the facts." Given this response, the farmers are likely to feel that they've wasted their time. A failure in identification has occurred. The semiotic chain has been broken because of a faulty interpretation or representation.

We could look at the problem as either a failure of the farmers' interpretive habits or a failure of the experts' representational habits. The farmers expect experts to tell them what to do. In terms of technical communication genres, they are expecting a manual or a proposal.

But the experts are giving a report, without going so far as to offer overt recommendations. At most, they may be recommending "further research"—the preferred activity of their discourse community. Their very refusal to come to conclusions or to offer recommendations may be an implicit promotion of a continuing research program.

Ultimately, we may view the failure to communicate as the fault of the scientists and as a problem of genre [1].[1] The text of the presentation takes the form of a report when its purpose should be either promotional or operational; the scientists should offer either a proposal, recommendation report, or a manual (how-to lecture).

The farmers are people of action; they want to know what the scientific study tells them about how to conduct farming in the region. Even though the scientists stake their image of professionalism on their claims to objectivity, which they feel is vitally connected to their reticence in promoting agendas of community action, they have accepted an advisory role in their acceptance of the speaking engagement, which demands that they address the needs of the community more or less on its own terms [2].[2]

Genre, Form, and Convention

Finding the right kind of representamen or the right collection of appropriate signs—the right code—means finding a kind of writing, speaking, illustrating, or acting that will best enable the identification of object and interpretant.

This semiotic problem mirrors the relationship between the scientists, who may be viewed roughly as a kind of macro-object, and the farmers, who may be viewed as a macro-interpretant. Clearly choices about genre, pre-coded collections of individual signs, or macro-representamen, are involved.

The diagram we used in Chapter 2 to show an engineer engaged in semiosis (reproduced in Figure 22) sheds some light on this case. If the engineer in this diagram has effectively analyzed the project objectives and has clearly imagined the actions that will result from the document, that writer is more likely to have produced effective verbal signs. But the success of the project outlined in the diagram depends upon the technicians' ability to perform the work after reading the manual. Writers are always at the mercy of the reading abilities and interpretive habits of the audience—though experience and audience analysis can help them shape the verbal signs in such a way as to determine more precisely how typical readers will react.

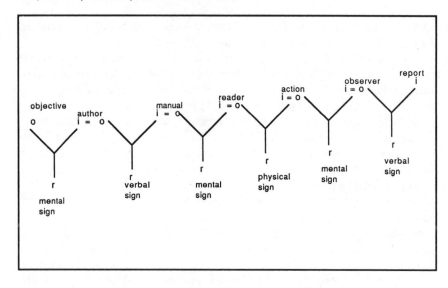

Figure 22. Expanded diagram of the semiosis of an
engineering project (reproduced).

To become effective, the interchange of documents and actions must feed back into itself. The original objectives look forward and backward in the semiotic series. In this process, *conventions* are formed, ways of thinking, writing, and acting that have been established by custom and practice over a period of time. Identification depends in part upon the writer's capability of satisfying audience expectations through adherence to convention and in part upon developing mutual interests in justified departures from convention.

A *genre*, in this sense, is a collection of signs that follows an expected or conventional *form*. As the rhetorical theorist Kenneth Burke suggests, form depends upon "an arousing and fulfillment of desires" [3, p. 157], usually the desires of a group of readers with particular interpretive habits.

The genre of the detective story, for example, assumes a reader who will identify with the investigating detective in seeking clues to solve the mystery. To satisfy the desire for interest and suspense, the author must permit the reader to come close to a solution without getting to it before the detective. If the reader gets the answer too far in advance of the detective/hero, the identification will be broken, and the reader will feel intellectually superior to the detective (and maybe to the author).

In our example of the soil scientists' presentation, the farmers expect the scientists to give them a proposal, even if they plan to reject it. They expect to be led from the facts to a set of possible actions. If authors have desires contrary to those of the readers, they must take conventional forms into consideration, even if they finally depart from them.

For example, the scientists may begin the talk by saying that the presentation of the facts will be followed *not* by a recommendation but by a discussion of possible alternatives with input from the audience. As always, the authors must acknowledge conventional practices of writing and reading either by writing within those conventions or by explaining why they are breaking the rules.

Genre Simplifies Reality

In developing a general theory of genres in technical communication, we will need to simplify practical reality a great deal [4].[3] All written genres—essays, novels, plays, poems, newspaper reports, and advertisements—give rise to subgenres. The subgenres of the poem, for example, include the lyric, the epic, and the ballad. The newspaper report may be a plain news story or a "feature."

The major genres of technical communication—the report, the manual, and the proposal—all have myriad subgenres. A manual, for example, may be (among other things) a user's guide or a maintenance book, a tutorial or a reference manual. But, in order to maintain the simplicity required to make our model broadly applicable, we will deal with each genre, at least initially, as a general umbrella for these different types of related documents. Genres are *generic.*

As we apply our model to an understanding of the genres of technical communication, we must keep in mind that, like all signs and combinations of signs, any report, manual, or proposal must coordinate the processes of writing, reading, and action. The projected outcome of a document—the action that is intended to result from the text—may get lost in either the production stage or the consumption stage of the communication process, in either writing or reading, representation or interpretation. Successful communication is a two-way street.

Genres, Modes, Formats: What's the Difference?

Since technical communication is primarily action-oriented, we can classify communicative strategies according to how they fit into the normal cycle of a technological project. There are at least three generic ways a document can help in developing a project:

* It can *report* on the project in its various stages.
* It can *promote* the project.
* It can help to implement and *operate* the technologies of the project.

Thus we can define at least three *modes* of technical communication:

* The reportorial mode
* The promotional mode
* The operational mode

A single document may contain several modes in its various parts. A technical proposal, for example, must *report* on the proposer's past experience in similar projects, and it must describe how the design will *operate*. But the promotional mode is dominant in the document as a whole. Proposers have their eyes on the future, and they gear their rhetorical strategies to convince others that their projections are credible and reasonable.

Formats—the way the pages of the document are mechanically arranged—may vary greatly regardless of shifts in mode and genre. A proposal may be either a single-page letter or a multi-volume book, depending upon the requirements and contexts of the project. Memos, letters, and full documents, if studied carefully, usually reveal that one of the three modes is dominant.

Each of the three modes is dominant in a particular *genre* of technical writing. The reportorial mode produces the report, including all of its subgenres—the research report, the progress report, the project report, the recommendation report, and so on. Despite this potential variety, all reports are a kind of narrative. Upon what might otherwise be ordinary stories, they impose the forms of narrative conditioned by the demands of practical reasoning. A research report, for example, tells the story of an experiment or a set of observations made by a researcher, but it mimics inductive logic by rearranging the narrative into the so-called IMRaD organization (Introduction, Methods, Results, and Discussion).

The promotional mode most commonly takes form in the proposal, though it can also be seen at work in such documents as technical sales pamphlets, annual reports, and even resumes and letters of application. The promotional mode deals with the future; its aim is to project and explain a solution to a problem. It must be highly persuasive in its rhetoric, for it usually has competition. Moreover, it deals with an uncertain future, in which facts do not yet exist. It thus requires the same kind of persuasive techniques included as part of the program of both classical and modern rhetoric [5-9].

The operational mode, represented in the various forms of the manual, deals with activities in the present by helping the reader to implement a project and to operate various technologies. It is clearly an appropriate ground for applications of pragmatic semiotics because of its emphasis on practice and outcome.[4] The dominant modes and genres of technical communication are summarized in Table 2.

Genre, in the simplest sense of the word, then, means a *kind* of writing recognized as distinctive by writers and readers. In a more complex sense, a genre is a set of conventions governing the arrangement, style, and general method of communication. These conventions are shaped by practice and are therefore in a constant state of evolution. In a theoretical insight that suggests that genres are, at the same time, makeshift paths to clear communication and actual foundations for communication, Charles Bazerman writes, "though genre emerges out of contexts, it becomes part of the context for future works" [10, p. 8].

Table 2. The Modes and Genres of Technical Communication

Mode	Orientation to Project	Genre	Time Conveyed	Purpose Primary	Secondary
promotional	prior to	proposal	future	to project, plan, and persuade	to explore
operational	during	manual	present	to enable performance	to explain and encourage
reportorial	during and after	report	past	to confirm, explain, and evaluate	to explore and recommend

GENRES AS CODIFIED RELATIONS BETWEEN WRITERS AND READERS

Writing in different genres, an author will experience shifts in relation to audience, subject matter, and text. Even the version of the self portrayed in the text will change. We can model these changes by using our semiotic theory. The view that results from this model suggests that *the genres represent convergences of various semiotic chains.*

The Report: Relating Past to Current Actions

With our triads greatly simplified, we can begin by modelling a typical rhetorical situation involving a technical report.

A writer has just finished some research and has on hand notes, tapes, and other "protoscribal" materials [11].[5] The writer works with these rough signs until a formal report is produced, which is then read by a supervisor or another researcher. The reader is essentially a judge of the report, who will determine whether the research—and the semiotic chain associated with it—should be continued in further actions. Figure 23 is an attempt to diagram the various relations.

To comprehend the concept of genre and thus to move to a discussion of rhetorical effectiveness, we must think of the report as existing at the juncture of two semiotic chains. The reader is also a researcher, who goes to the literature as part of the normal research process or to help get past a stumbling point in research. In the literature, this reader, the other researcher, encounters the original report. Figure 24 shows this relationship. The process depicted here is a common step in research—usually an early step, the "literature review." The report literature—whether in science, business, or government—is an effective means of

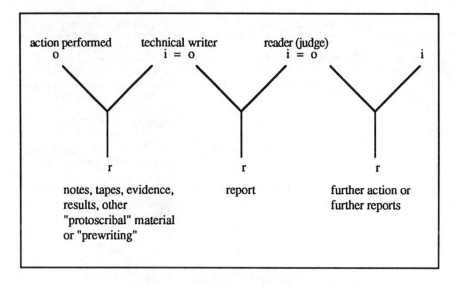

Figure 23. Report semiosis.

linking the work of various agents and therefore facilitating or making possible further actions.

Many of the rhetorical requirements of the report genre begin to make sense in this light. Consider, for example, the reasons for using conventional formats and stylistic patterns, like the IMRaD organization. The purpose of such guidelines is to smooth the reader's path to action by putting information into a form that satisfies the reader's expectations and that can be rapidly consumed and converted to action.

The Manual: Reconciling Producers and Users

Now for the manual. Figure 25 is a diagram of a common kind of situation, in which a technical writer develops a user manual to help office workers use a computerized spread sheet.

From the perspective of the office worker, the user manual is a means of getting a job done—in this case, a necessary set of instructions for using a spread sheet program to organize a set of new accounts passed along by management. From the perspective of the technical expert who developed the spread sheet software, by contrast, the user manual is a representation of the product.

While a rhetoric emphasizing the perspective of the technical expert would thus be *product-oriented,* a rhetoric emphasizing the perspective of the user would tend to be *person-oriented.* The technical writer, however, must consider both perspectives and must mediate between accurate representation of the product, on the one hand, and the needs of the user, on the other.[6]

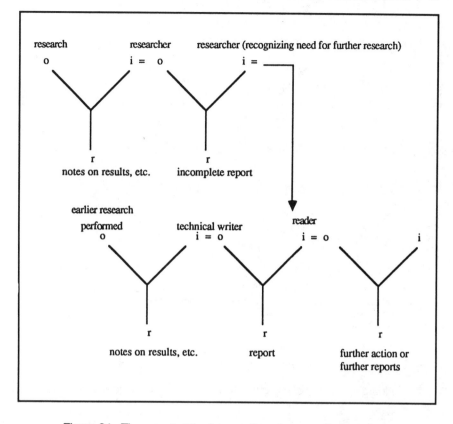

Figure 24. The report at the intersection of two semiotic chains.

A pragmatic emphasis on outcomes, on appropriate and effective actions, is the link between the various perspectives. Everyone has an interest in that outcome, an action which depends upon an effective manual. The user is able to perform the desired action, while the technical expert may enjoy the success of a product that works.

Thus, as the rhetorical theorist Carolyn Miller has rightly argued, *genre itself is a form of social action* [1]. In the case of the manual, the social action involves building communities among various technological interests—software engineers, technical writers, business managers, and office technicians.

The Proposal: Building Bridges

This bridge-building through textual action is perhaps even clearer in the proposal genre (as Sanders has noted [5]). We can alter the case in the manual example to show what we mean.

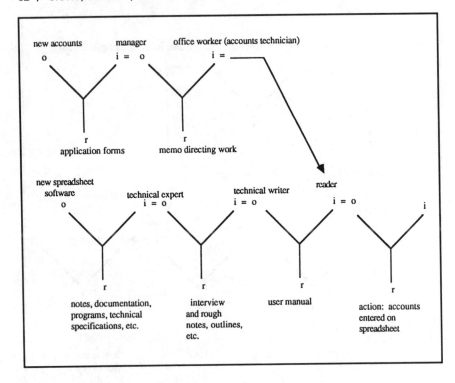

Figure 25. The manual at the intersection of two semiotic chains.

An accounts manager, let's say, has a need for a particular kind of custom software. The need arises from the manager's recognition that the current accounting technology is inadequate for the special purposes of the company. The manager has a Request for Proposal (RFP) written and published in the *Commerce Business Daily*, where it is read by a technical writer who manages the proposal efforts of a software development company.

The writer then collaborates with a technical expert on a proposal that puts forward the company's new prototype spread sheet as an appropriate fulfillment of the needs outlined in the RFP. The accounts manager reads the proposal and offers a contract to the software company to develop the custom software. Figure 26 is a diagram of the situation.

The differences among the diagrams in Figures 24, 25, and 26 hint at the major rhetorical differences among the technical genres, differences that we will elaborate further in Chapters 5 and 6.

But we should also recognize some important similarities. Above all, the genres share a typical outcome. All of them end in a relationship something like the one diagramed in Figure 27.

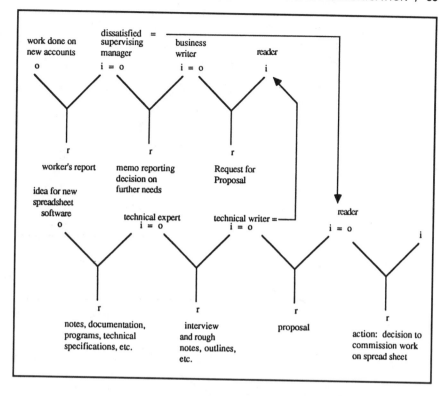

Figure 26. The proposal at the intersection of two semiotic chains.

This relationship is a version of what scholars in composition studies have come to think of as the standard "rhetorical situation." The main difference between this and the rhetorical situation as outlined by scholars working exclusively within an academic context (especially those who focus their work on freshman English) is that the situation of technical communication is pragmatically oriented toward a resulting action rather than being oriented toward the production of further discourse (see Chapter 8).

MAPPING GENERIC DIFFERENCES BY USING RHETORICAL PRINCIPLES

In an earlier paper [12], we used an even simpler diagram to show the relation of writer and reader to text and action in technical communication. This version of the diagram, shown in Figure 28, is an adequate summary of the relationships suggested by the more complex triadic diagrams, which we can use to map the differences among the genres.

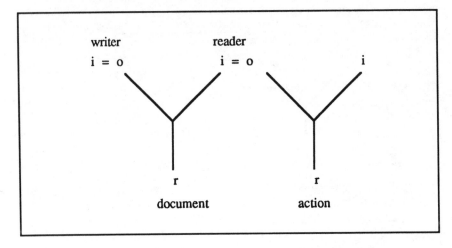

Figure 27. The outcome of the semiosis of technical documents.

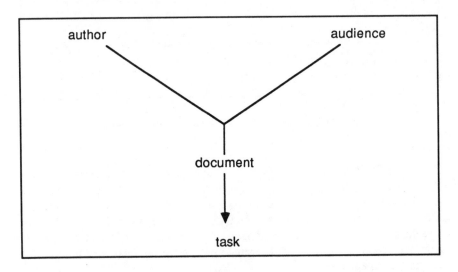

Figure 28. The rhetorical situation of technical documents.

The relationship of author and audience—the chief concern in the field of practical rhetoric—is the key to the differences among the technical genres. Every technical document involves an exchange of knowledge and/or power between an author and an audience.

The software engineer in the case diagrammed in Figure 25, for instance, has the knowledge and the power to analyze business accounts by using a new computer program, but has no personal use for this knowledge. The technical writer has the knowledge and the power to communicate the technical expertise of the engineer to a nontechnical readership. The accountant needs the knowledge of both the technical expert and the technical writer to be able to use the new technology to do the work of the office. The three agents collaborate in an exchange of expertise so that all benefit.

This knowledge/power exchange differs from genre to genre, in ways that the full diagrams (Figures 24-26) make clear. The degree of mutual dependence of author and audience shifts as we move from genre to genre, as does the quality and kind of relationship between author and audience. We can use the simple diagram of the rhetorical situation (Figure 28) to map and briefly explore these differences.

Author-Audience Relations in the Report

A report presents the author's knowledge about and evaluation of a project to an audience of readers that will judge the results and make decisions about future actions. These actions may involve either the author or the audience. The rudiments of this relation are shown in Figure 29.

The degree of the author's interest in the eventual action influences how much new power this author achieves in passing knowledge to the audience. The

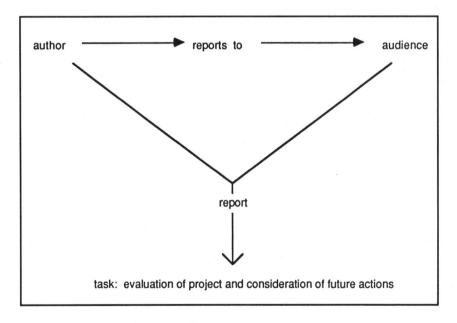

Figure 29. Map of report rhetoric.

audience's degree of dependence also rests upon the author's interest in further actions. If the author is an employee or contractor for the company owned or managed by the audience, the power balance is clearly different from a situation in which the author is an outsider hired as an expert consultant.

Author-Audience Relations in the Manual

A manual enables the audience to grasp the knowledge of the author and thereby to put a product or a system into operation, as shown in Figure 30.

The power relation is somewhat clearer here. The reader is nearly totally dependent upon the author to provide accurate and usable information. The author, who is rarely involved directly in the consequent action, must recognize, however, that, having paid for the information that will enable the action, the dependent reader can be the most demanding of all audiences.

Author-Audience Relations in the Proposal

A proposal promotes a project to a reader who will award funding or permission and will thus enable the proposer to act, as shown in Figure 31.

This is perhaps the most complex of the knowledge/power relations. In the case of a contract proposal, for instance, the audience cannot act without the help of the

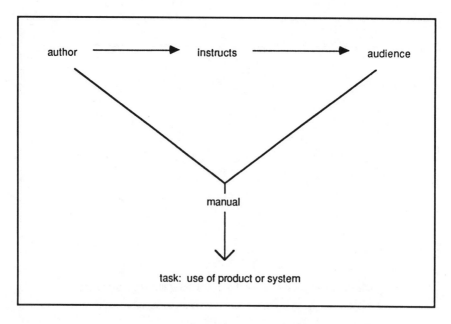

Figure 30. Map of manual rhetoric.

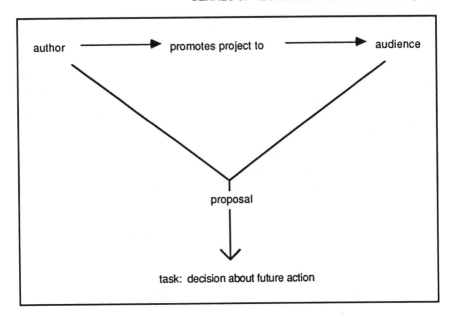

Figure 31. Map of proposal rhetoric.

proposer; the RFP is essentially a call for help [13]. But the provider of that help is strongly interested in the outcome of the document. Moreover, the author must compete with other authors to become involved in the ultimate action. Compared with the authors of other genres, the proposal author has the highest degree of interest in the success of the document, success which is measured in that author's ultimate participation in extending the project.

Genres: Codes for Communal Actions

With these rudimentary maps in place, we can now explore more fully a couple of dominant themes in the rhetorically oriented studies of technical communication. Chapter 5 will focus on audience; Chapter 6, on author. In both of these chapters, we will argue that an effective treatment of these themes requires an approach through genre theory, which in turn rests upon the kind of socially oriented semiotics that we have begun to develop in the last two chapters.

Genres are shorthand codes for describing typical kinds of communal actions. When writers say, "I am writing a proposal," or "I am writing a report," or "I am writing a manual," they are not just describing the kinds of words and pictures they are putting on paper; they are describing their current status in a power/knowledge complex.

NOTES

[1]Or we may view it as a problem arising from a conflict of discourse communities. See Chapter 8. But also realize that the two theoretical themes, genre and community, are intimately related, as Carolyn Miller has shown in her excellent article "Genre and Social Action" [1] and as we argue later in this chapter.

[2]For a fuller treatment of the role of the scientific expert in public debate, see Killingsworth and Steffens [2].

[3]Remember the words of Kenneth Pike: "Only if a theory is simpler than that reality which it is in part reflecting is it useful" [4, p. 6].

[4]Recent work in the rhetoric of manuals has also taken a rather unexpected turn toward mentalism, with articles in both theory and practice advocating a newly humanized tone. See Chapter 6 for a discussion of the literature on the use of "personae" in manual writing.

[5]The term, and so far as we know the concept of, "protoscribal" materials comes from the work of Robert Zoellner [11]. Zoellner coined the term to describe the kind of signs that mediate talk and writing (or thinking and writing).

[6]The position of the technical writer is a politically and ethically sensitive one for this reason. The ideological factors influencing this relationship of technical expert, technical writer, and nontechnical reader form a major concern in Part III.

REFERENCES

1. C. R. Miller, Genre as Social Action, *Quarterly Journal of Speech, 70,* pp. 151-167, 1984.
2. M. J. Killingsworth and D. Steffens, Effectiveness in the Environmental Impact Statement: A Study in Public Rhetoric, *Written Communication, 6,* pp. 155-180, 1989.
3. K. Burke, *Counter-Statement,* Harcourt, New York, 1931.
4. K. Pike, *Linguistic Concepts: An Introduction to Tagmemics,* University of Nebraska Press, Lincoln, 1982.
5. S. P. Sanders, How Can Technical Writing Be Persuasive? *Solving Problems in Technical Writing,* L. Beene and P. White (eds.), Oxford University Press, New York, pp. 55-78, 1988.
6. C. Kallendorf and C. Kallendorf, Aristotle and the Ethics of Business Communication, *Iowa State Journal of Business and Technical Communication, 3,* pp. 54-69, 1989.
7. M. J. Killingsworth and M. K. Gilbertson, Rhetoric and Relevance in Technical Writing, *Journal of Technical Writing and Communication, 16,* pp. 287-297, 1986.
8. M. J. Killingsworth, M. K. Gilbertson, and J. Chew, Amplification in Technical Manuals: Theory and Practice, *Journal of Technical Writing and Communication, 19,* pp. 13-29, 1989.
9. M. K. Gilbertson and M. J. Killingsworth, Classical Rhetoric and Technical Communication, *Teaching the History and Rhetoric of Scientific and Technical Literature,* S. Southard (ed.), Association of Teachers of Technical Writing Anthology Series, ATTW, forthcoming.
10. C. Bazerman, *Shaping Written Knowledge: The Genre and Activity of the Experimental Article in Science,* University of Wisconsin Press, Madison, 1988.

11. R. Zoellner, Talk-Write: A Behavioral Pedagogy for Composition, *College English,* *30,* pp. 267-320, 1969.
12. M. J. Killingsworth, Toward a Rhetoric of Technological Action, *Proceedings of the International Technical Communication Conference, 34,* pp. RET-136-139, 1987.
13. H. Holtz and T. Schmidt, *The Winning Proposal—How to Write It,* McGraw-Hill, New York, 1981.

CHAPTER 5

Generic Audiences in Technical Communication

THE GENERIC APPROACH TO AUDIENCE

At least since the late 1960s, when Thomas Pearsall [1] and other rhetorically inclined scholars effectively introduced the long tradition of audience awareness to technical communicators, questions about audience have dominated theoretical and practical commentary on the planning and writing of technical documents. Commentators have introduced various methods and theories on the needs, habits, psychology, and sociology of technical readers.

The breadth and depth of the early literature has been gauged in survey articles by Keene and Barnes-Ostrander [2], Sanders [3], and Masse and Benz [4], while Jo Allen has surveyed new directions [5]. Our task in this chapter is to reduce the advice in the literature to a few general theoretical positions, which we can evaluate and incorporate into our theory of technical genres.

The basic position that emerges is this: Though audience cannot be ignored, it must be understood not as an isolated theoretical entity, but in its relationship to an author with a definite purpose for writing. Moreover, the concepts of author and audience are radically limited by considerations of *genre*.

In fact, most commentators in the field already appear to subscribe to this viewpoint, though they may not consciously recognize their shared theoretical ground because they rely on various methods of how to approach the readership of a document. Their focus on traditional questions of audience obscures their kinship in theory.

Our classification of genres provides a useful way of showing that different kinds of documents give rise to different coded relationships between writer and reader. These relationships tend to appear again and again regardless of what method drives audience analysis.

Cognitive or Cultural?

In the bulk of the literature (old and new), audience awareness is approached from one of the two perspectives outlined in Chapter 3, the cognitive or the cultural.

The cognitive approach tends to assume a *universal audience*—or at least a *unified audience*, the busy, fast-reading problem solver so common in corporate and public life. Bolstered by experimental and laboratory studies, which treat subjects as universal or typical readers, the cognitive approach to audience awareness emphasizes points of general readability and usability (see, for example, Barker's collection of essays on software documentation, especially the introduction [6]).

By contrast, the cultural approach portrays an audience conditioned by the particular circumstances of each communication situation. Insisting on naturalistic study—on audience analysis that considers actual users in actual contexts of use—cultural analysts owe their greatest debt to media studies and market research. Discourse community theory grows out of this approach (see, for example, Matalene [7] and Part III).

For the culturalists, audience analysis is truly analytical. To determine how specific readers react to a publication, culturalists seek out information about demographic factors on readers, such as background, age, occupation, and familiarity with the subject matter [5, p. 53]. The proliferation of literature on usability studies in the last decade extends this tradition, showing how methodologies pioneered in ethnographic investigations and even in cognitive science may be applied to the study of representative audiences and corporate cultures [8].

As we suggested in Chapter 3, these two approaches are reconciled as two sides of the same question in a pragmatic theory of signs. Both cognitive and cultural approaches to audience analysis may indeed be necessary, since one (the cultural) will provide specific information about representative users, while the other (the cognitive) will give writers ways of accommodating readers overlooked in the market study or usability test.

The cultural approach thus generates an image of an actual audience; the cognitive, of an ideal audience. Each perspective therefore complements the other and compensates for its deficiencies.

However, a third position has recently arisen that calls the whole enterprise of audience analysis into question.

Audience or Author?

This third position questions the very efficacy of audience analysis. Going against the grain of the dominant audience-centered theories—such as Pearsall's, which is all but "writer-effacing" [3, p. 58]—these newer studies tend to be author-centered. They suggest that the audience becomes inconsequential in the mind of the writer during the process of writing.

According to this critique, theorists "violate the intentions of original authors" by overemphasizing the need for audience analysis; thus, instead of analyzing and addressing an audience, writers should "construct roles that flexible readers may choose to adopt" or choose to reject [4, p. 53]. Not surprisingly, many author-centered theorists come from the ranks of composition studies [9, 10]. University teachers must be concerned with helping students to develop an authorial (and authoritative) stance.

But even a writer as experienced in industrial technical communication as David Dobrin has adopted an author-centered position. In a phenomenological critique of writing practice, Dobrin has suggested that authors typically forget the audience and write with the subject matter of the communication foremost in their minds, giving it the best possible form [11]. The struggle of writer with subject matter, he argues, will naturally produce the clearest and most effective document.

Genre—A Mediating Position

The author-centered position, which reveals some interesting holes in the audience analysis theory, is itself put into question by a consideration of genre. Genre theory suggests that, in the very act of giving form to subject matter, writers must deal with audience—maybe in spite of themselves.

This encounter with the generic reader may well be unconscious. Authors are able to forget audience expectations because these expectations are so thoroughly coded into the genre that governs the writing. A proposal audience, for example, is an evaluator, a judge, someone who holds decision power over the author.

Dobrin's ideal author is more or less resigned to ideal audiences of this kind, audiences provided by the generic codes. This attitude of resignation could be damaging if the author's purpose is to *change* the audience or otherwise to *affect the habitual actions* of an audience. We think therefore that it is unwise to ignore the audience in technical communication.

But it is also important to recognize that what often passes for a concern with audience may be better understood as a concern for genre—the textual patterns by which authors relate to audiences in a social setting. Because of the tendency of genre to intervene in the habits of both authors and audiences, genre becomes a mediating force between the theorists who insist on audience analysis and those who emphasize authorial stance. Much as the pragmatic concept of the sign mediates the dispute between cognitive and cultural approaches to document design, the concept of genre thus provides a way of focusing not on either author or audience but instead on the *relation* of authors to audiences.

The approach through genre takes account of the critique offered by each side against the other. It allows that writing cannot simply cater or pander to the audience and also that writing cannot merely serve as the vehicle of the author's expression. Instead, it insists that the experience of communication changes both parties—writers and readers—by bringing them into contact with each other.

An interest in genre animates much of the advice in the older literature on audience analysis.[1] In committing to a particular genre of writing, a writer also commits to a particular set of audience expectations. And vice-versa, in committing to a particular audience (or audiences), the writer tends to favor a narrow range of genres. From researchers, funding agencies expect grant proposals. From teachers, students expect textbooks. From the computer company, users want manuals. From consultants, companies solicit feasibility studies. Genres frame expectations and place writers and readers into easily identifiable social roles.

Though audience expectations differ from genre to genre, it is possible, through the study of typical contexts, to arrive at general principles about what the reader expects from a text and what kinds of texts readers expect for different kinds of information.

Generic Readers

In his early work, Pearsall [1, 12] was already moving in this direction. He claimed that technical writers usually address one of several kinds of generic readers—technicians or executives, laypersons or experts [13].[2] Technicians want detailed data and careful descriptions; executives want recommendations, conclusions, and costs. Laypersons want explanations; experts want high informational value and density.

An author's own relation to the subject matter cannot be a reliable rhetorical guide, Pearsall and other rhetoricians argue, because it may lead the writer to ignore audiences that do not share the writer's interests, educational background, or general orientation toward action. A technician writing for an executive, for example, is all too tempted to provide more details about mechanisms and processes than the reader cares about while skimping on the recommendations and bottom-line funding information that the reader wants.

The full generic approach to audience extends this outlook. A genre provides a flexible code that allows the writer to address multiple types of audiences within the same document. A technical report, for instance, allows the writer to accommodate both an executive readership (in the executive summary) and an audience of technicians (in the technical discussion). As Robert Hodge and Gunther Kress have found in their work on social semiotics, genres are "typical forms of text which link kinds of producer, consumer, topic, medium and occasion" and which thereby "control the behavior of producers of such texts, and the expectations of consumers" [14, p. 7].

When writers analyze their tasks in generic terms, they discover principles that help them to determine which information is *relevant* to their readers' needs. And, as Hodge and Kress indicate, these principles relate not only to questions about author, audience, and medium—the major concerns in the literature on audience analysis—but also about the *occasion* for writing, a topic which introduces questions about *timeliness*.

Relevance, Occasion, Timeliness

Most technical communicators will agree that relevance is a key concept. In judging the success of a particular document, questions of relevance invariably come into play: How does the report meet the reader's current informational needs? How well is the manual suited to the user's abilities at the beginning of the learning process? How well does the proposal show the interrelation of the proposer's present needs with those of the funding agency?

At the heart of the concept of relevance is the element of time. When something is relevant, it is pertinent in the present. The report, in other words, must show how its data are related to the reader's present and future plans. The manual must help the user with the problem at hand. The proposal must demonstrate a current understanding of the conditions that make a proposed project necessary and must establish present trends that justify the proposer's future plans.

A relevant text is thus present-oriented, but fully aware of the semiotic chains converging from the past and extending into the future. It evokes in the reader a feeling of a meaning-full and opportunity-saturated present by placing the reader at the center of newly joined discourse paths.

Many of the clichés associated with good writing in general apply to the relevant text in particular, because they connote the necessity of time consciousness. To "make events come alive" is to endow them with an aura of the present. To "involve readers" means, in one sense, to obliterate their sense of real time and enwrap them in the "eternal present" of the discursive experience. And, among book reviewers, no higher compliment may be paid to a piece of nonfiction writing than to say that it is "timely."

Relevance is particularly important in technical communication, which takes place in a world of action, a present-oriented world where the chief purpose of writing is to facilitate purposeful conduct. As we suggested in Chapter 4, the documents that shape and reflect this activity indicate various stages of a project: The plans appear in the proposal; the procedures appear in the manual; the results are evaluated in the reports.

But the orientation of a document to the project as a whole is, *from the perspective of a reader at any moment in the project,* not as important as the need to relate the information to the immediate present. Although the proposal deals with future actions, the manual with actions as they are undertaken, and the report with past actions, *all the genres must create the impression of a profoundly present reality.*

Natural Time and Rhetorical Time

Among the concerns of the writer, then, is a resolution of a potential conflict between what we call *natural time* and *rhetorical time*. Natural time is the orientation of the document to the overall project; it represents the writer's present. Rhetorical time, by contrast, is always the *reader's* present—the time at which the document is in use.

Each of the genres conveys the sense of urgency or immediacy implied in the concept of rhetorical time through different conventions for the use of grammar, style, graphics, organization, and tone. The genres create different combinations of the classical "appeals" to an audience—*ethos*, or the appeal to the audience's sense of the author's character; *pathos*, or the appeal to the emotions; and *logos*, the appeal to reason [13].

In the remaining sections of this chapter, we will briefly show some of the particular rhetorical needs of each genre in establishing relevance. Along the way, we will demonstrate how the generic approach, covertly if not overtly, informs the traditional concern with audience.

THE REPORT

In their well-known historiographical work, *The Modern Researcher*, Barzun and Graff note that the "great difference between the [historical] scholar's main interest and that of the ordinary report-writer" is that "The former seeks to know the past" while "the latter is concerned with the present, generally with a view to plotting the future" [15, p. 6].

This distinction between knowing and doing corresponds to the distinction that the rhetorical theorist Walter Beale makes between scientific or "contemplative" discourse on the one hand and technical or "instrumental" discourse on the other hand [16]. In technical report writing, increases in knowledge and information or alterations in consciousness are relevant only if they promote improvements in action. Moreover, *to know* something implies an awareness of the past, whereas *to act* suggests an orientation in the present.

Grammar

If nothing else, this distinction suggests an appropriate verb grammar in report writing. Historians, at least narrative historians, may restrict their writing to the simple past tense (often called the "historical past"). But the technical reporter, in helping the audience relate the past to the present and the future, must resort to the full range of aspects of the past tense available in English.

The present perfect, for example, presents actions that occurred in the past but continue to operate in the present: "The researchers *have begun* the work for the initial phases of the project," we might write in a progress report, instead of "The researchers *began* the work." The perfect tenses in general, as well as the progressive tenses (they *are working*, we *were working*), tend to organize actions according to their exact relationship in time, thus freeing the simple present to express only habitual actions ("I work ten hours a day") and the simple past to convey history ("They worked hard all their lives").

Style

The grammar of tenses, however, is too governed by habit to evoke fully the feeling of present reality in the reader. Thus, in the case of the report, the writer must find new ways to bring the past to bear on the present. Stylistic choices may have a stronger impact in achieving this goal.

The use of the active voice rather than the passive voice sentences, for example, aids temporal visualization by making the context of the writing more specific.[3] Consider the following two sentences:

1. The experiment was successfully completed.
2. Our researchers successfully completed the experiment.

The second sentence involves more than the switch from passive to active construction. The people responsible for the key action are now the grammatical subjects of the sentence. According to the rhetorical critique of the impersonal, passive style, the first sentence is inferior to the second because, by removing the actors, it eliminates the drama of the narration; it takes the story-telling out of the report. The second sentence takes advantage of the inherent strength of the narrative mode, the "human interest" appeal—the pathos—which the rhetorical scholar Frank D'Angelo characterizes as "intensely dramatic and dynamic" [17, p. 212].

D'Angelo advances a strongly rhetorical view of relevance in his use of the term "dramatic" to describe this stylistic ideal which places human agents in the subject position and uses active verbs wherever possible. Drama is a genre of literature that always presents action in the present. The audience of drama must focus on the action one moment at a time. Though it may be set in the past or the future, the success of a play depends on its bringing actions to life as a present reality. A "dramatic" style, then, is one that accomplishes the goals of a rhetoric of relevance.

Consider two more sentences:

1. The Russians are ahead of the United States in developing technology for transport vehicles.
2. The Russians have been ahead and continue to lead in the race to develop technology for transport vehicles.

The substitution of the present perfect and the emphatic compounding of the verb in sentence 2, as well as the use of the temporally suggestive noun *race*, enhance the reader's consciousness of time. The author conveys an attitude of urgency and an encouragement to action. An implicit recommendation and appeal, perhaps even a judgment, replace the bland objectivity of sentence 1.

Striving for objectivity, as represented in the excessive use of the passive and impersonal style, also leads writers to avoid such uses of judgment. In scientific writing for peer review, a different kind of judgment—one involving interpretation of data and directions for research—may well be expected. We will consider further these stylistic differences in Chapter 7. For now, the important point about report rhetoric is that judgment and the stylistic markers thereof nearly always find a place in action-oriented writing, no matter how "objective" the tone.

Tone

Authors often have more flexibility than they realize in the cultivation of tone—the means by which they reveal their attitudes toward their subject matter and audience, their *ethos*, in Aristotelian terms. Citing his experience as a consultant, Thomas Pearsall [18] has noted that what is often condemned as "bad writing" is in reality writing that, though grammatically correct and stylistically acceptable, fails to rise to a level where readers can evaluate the findings reported. Facts like "the Memphis office regularly ignores Standard Accounting Procedure 501" roam about in a report meaninglessly, because they provide insufficient information on which the reader might make meaningful decisions about the data. Relevance depends upon evaluation and judgment, which depend in turn upon an awareness of the typical audience's needs.

In fact we may define *judgment* as the ability to make information relevant to action. The audience's response to bald presentations of fact is often, "So what?" or "How is this relevant to our actions at the present time, and what does it say about future actions? Have the accountants at the Memphis office, in failing to follow Standard Operating Procedures, irreparably damaged the company, or have they landed upon a brilliant innovation that cuts through bureaucratic red tape?" [19][4]. Just as academic writers are supposed to show how their theoretical meanderings are relevant to the "real world," writers in business and government must structure their presentations of factual information in a way that reveals relevant judgments.[5]

Mere efforts to evaluate data, however, are not sufficient to convince an audience of the relevance of the data. Writers and speakers must convince their audience of their own competence to evaluate within the context of shared values. This is best accomplished by making clear the chain of logic the presentation follows: What factors led to a negative or positive evaluation? What was the basis for judgment?

Writers must therefore stress the assumptions on which they base their reasoning (for example, an action may be considered good if it increases efficiency). By stating what might seem obvious, writers show their good character—their *ethos*—by indicating that they hold certain values in common with their audience of employers (or, by indicating a divergence of values, they give the audience a chance to make appropriate adjustments). In short, logos and ethos provide paths for demonstrating professional competence.[6]

Graphics and Interpretation of Data

Technical reporters have found, in analyzing audiences, that one of the most effective means of providing clear judgments through logical structure is to use graphics that are properly integrated with written text. In reports, comparative graphics built around an x-y axis are particularly useful, especially in the division and classification of data, a prime requisite for judgment.

One semiologist of graphics, Jacques Bertin, argues that "visual perception is spatial perception that allows anyone to use a new system of classing: the simultaneous consideration of different elements" [20, p. 7]. The information contained in a report is not merely individual items of data but is rather the *relationship* among elements, subsets, or sets [20, pp. 12-13]. To discern such relationships is to take a crucial step toward making a good decision.

As an example, consider Table 3, which gives the number of student users for three different computer labs over a period of one year. According to Bertin's system of analysis, a decision-maker can use this table (or any display of data along an x/y axis) to develop three levels of interpretation:

1. *The elementary level.* Every entry along each axis represents a single element as it is placed in relation to other single elements. At this level, for example, the table shows that forty-seven students used Microlab 1 in January. Simple as this is, it reveals the power of the table as a rhetorical device for arrangement: Imagine the amount of verbal text that would be required to state all of the elementary relationships in the table.

2. *The intermediate level.* At this level, readers of the report, guided by the author's verbal text or through their own exploration of the data, can

Table 3. Number of Student Users in University Computer Labs (Example)

	Jan	Feb	Mar	Apr	May	Je	Jl	Aug	Sept	Oct	Nov	Dec
Micro-lab 1	47	22	25	31	30	35	31	33	59	44	43	50
Micro-lab 2	52	19	24	40	38	26	26	24	55	43	40	51
Main-frame lab	106	54	44	44	40	67	30	21	98	30	32	31

explore the relationship within subsets such as January, February, and March, and answer questions like "Which lab is most used in February?"

3. *The overall information level.* Decision-making is most often reserved for this level, since it offers answers to questions about overall trends in the relationship of x and y, such as "After an initial introduction, are students more likely to make heavier use of the microlabs or the mainframe lab?" Answers to such questions will guide policy decisions about funding and staffing the labs.

Bertin concludes that "a construction which does not enable us to define groupings in x and y does not reach the overall information level of the entire set. This is the mark of inefficiency" [20, p. 13]. The implication is that efficiency in decision-making depends on the kinds of groupings that graphics provide and that verbal text can, at best, weakly imitate.

In addition to efficiency, however, the report-writer is concerned with relevance, and, if the data are to be more than a collection of facts in the past tense—which, to our thinking, is always the "tense" of a table—then the overall information level can only be attained in combination with a verbal commentary. The commonplace of technical writing textbooks—that tables and charts should always be accompanied by explanatory text—attains new meaning in light of this principle. The author must guide the reader by posing relevant questions and explaining the meaning of the table for present and future actions. Evaluations, judgments, and decisions require such *contextualization.*[7]

THE MANUAL

Orienting information to the present would seem to be less of a problem in the manual than in the report, since manual writers design their documents specifically to help readers solve problems in the present. In this sense, the manual, not the report, is the central genre of technological or "instrumental" discourse [8].

Too often, though, the conditions under which technical manuals are produced distort the writer's own time consciousness to such a degree that both comprehending the user's point of view and conveying a sense of immediacy become difficult. More and more these days technical manuals are written during the development of a product or even before development [21-23], but still they are most often produced after the product is ready, or nearly ready, for the market.

The manual writer quite rightly thinks of the product as a completed entity and thus is led to describe it rather than to bring its capabilities to life as a present, working reality. The writer is led, that is, to write in the reportorial mode rather than in the operational mode. This approach may be appropriate for introductions or product overviews, but it is likely to fail in the instructional portions of a manual where the reader's point of view must be not only acknowledged but accepted as a leading structural principle (pathos).

Rhetorical study has offered a number of ways of building identification with the reader and representing that identification in styles. Morris Dean of IBM puts it this way: "writing from the reader's point of view brings the reader into 'the story,' so that it is easy for the reader to imagine what you are telling" [24, p. 2]. If "the story" is about the product's development, it is a thing of the past in which the reader/user can in no way be involved. If, on the other hand, the manual's "story" deals with the product's use in sequentially presented and explained steps, it occurs in the present and involves the reader directly.

The connection with drama, which in the case of the report could be construed only as an ideal of narrative, can be more immediately realized in the manual. The manual is an elaborate set of stage directions in a complex dramatic performance. The lead actor is not the product—it is a dead thing, a tool meant for human hands—but the user who has the ability to bring the product to life. Thus Dean can advise technical writers to "make the reader the hero of your 'how to' story."

Grammar

The manual, most writers agree, should be addressed to "you," the reader. This creates a relatively simple imperative grammar that allows the present tense to dominate the presentation.

It varies only when readers require explanation. But, even when the writer must stray from the imperative ("Connect wire x to plug y," or "press the enter key"), the focus on the user (pathos) is usually maintained through the use of the second person pronoun: "By using this connection, you avoid overloading the system," or "If you do not enter the data in this way, you risk having to re-format the table."

Organization

Focus on the present is also achieved through what has come to be known as "task-oriented" organization. Each section of the manual should present a unique task, a single operation. It should be more or less self-sufficient and should exclude features unrelated to the task. Moreover, the organization should reflect the order of use, and titles and headings should reveal the task at hand—"Setting the Margins," for example, rather than "Capabilities of the Margin Function."

The pace of presentation is a major consideration, as is the timing associated with introducing and defining new terms for the user. Writers carefully divide the manual, according to the needs of the user, into separate sections for learning tasks (the tutorial section) and recalling tasks or pieces of whole processes (the reference section).

So the logos of organization follows the pathos of the reader. For example, a common approach in task-oriented manuals is to organize the chapters so that the reader moves from the simplest, most frequently used information to the most complex, least frequently used information [25, pp. 16-17]. Relevance demands that writers emphasize the information that allows the most effective and immediate performance of a given task.

Tone

Manual users form a notoriously difficult audience. Since they usually rely on the manual only when action has failed ("When all else fails, read the instructions"), they take up the task of reading unwillingly and with frustration.

To anticipate and counteract the surliness of their readership, many manual writers strive for a sense of immediacy and a conciliatory tone by cultivating an illusion of character (ethos or persona) in their writing. As William Skees advises, "Look upon the user manual as the document which substitutes for your own physical presence, providing everything in the way of guidance and assistance that you, yourself, would provide to the user of the system if you were there" [26, p. 181; 27].[8]

"Physical presence" implies present time as well as immediate location, so that time consciousness again plays a part in manual rhetoric. The helpful voice of the manual is there when the reader calls. (The 1-800 numbers in the backs of many manuals take this approach to communication one step further, placing a helpful human voice at the other end of the telephone line, thus relieving the user of having to read the manual at all.)

The stylistic development of a personality, or a "persona," is in fact an extremely problematic matter in manuals, a matter we will treat in detail in Chapter 6. For now, we only note in passing that the conversational style is one way of constructing the relationship of author and audience—a path charted by the generic demands of the technical manual.

THE PROPOSAL

Writers of reports and manuals take information developed in the past and make it relevant for current tasks, while proposal writers take current knowledge and cast it into plans for the future. Like the science fiction writer, the proposal writer must establish credibility (ethos), which is part of being relevant since a credible text creates characters, problems, ideas, and facts that are immediately meaningful to the reader's sense of reality. In proposal writing, such materials have, or appear to have, the tangibility of past and present reality, a definiteness that offsets the uncertainty of the future. Probably for this reason, proposals for projects that are already well under way have a high rate of success.

Nevertheless, not all proposal writers have the luxury of working on projects begun prior to funding. They must develop a rhetoric for an audience of evaluators, a rhetoric that creates interest and desire, that establishes credibility through specificity, and that demonstrates the soundness of the proposal [28].

Grammar and Style

Since the proposal must show the relation of present interests to past trends and future possibilities, the use of the present tense, which indicates habitual action, and the future tense must, as in the report, be fully integrated with perfect and progressive tenses that show aspects of overlapping time and action sequences.

A variety of personal pronouns, which may seem inappropriate in other genres of technical writing, will appear frequently in proposals. Even in scientific proposals, *I* and *we* are common, and so is *you* on occasion, since the purpose of a proposal is to respond to a call for help, the Request for Proposal, or RFP. (Chapter 6 will take these matters up in detail.)

Tone

The tone a proposal writer should aim for, according to most rhetoricians in the field, is well expressed by Holtz and Schmidt's term *responsiveness* [29]. Since proposal readers are not likely even to read proposals that do not respond directly to their needs, proposal writers must pay particular attention to requirements stated in the RFP or, in the case of unsolicited proposals, the funding agency's statement about its particular mission.

Responsiveness is closely akin to relevance, especially in the implication of *timeliness*. A timely proposal responds to needs and problems that any proposal evaluator reads about daily in the professional quarterlies, the trade magazines, and the *The Wall Street Journal.*

But proposal writers are well advised to study the recent periodicals carefully, so that they can walk the fine line between timeliness and *trendiness.* The latter attribute is often the subject of warnings in statements by funding organizations. ("I can't bear to read another proposal about using computers in the composition class," one proposal referee told us.)

Specificity

Agencies that fund proposals usually demand specificity. For years, nebulousness has been a frequently cited cause of proposal failures [30, 31]. Proposals must be specific. By giving specific budget figures, for instance, the proposal writer provides a clear view of future needs, especially if such figures are shown to be based on current prices with adjustments based on current trends.

To borrow terms from historical geology, evaluators in business and government tend to take a uniformitarian rather than a catastrophist position on economics. They perceive time as an even flow rarely disturbed by shocking discontinuities.

The proposal writer is not expected to account for possible political revolutions or even stock market crashes, though these events are not uncommon in modern life. If in geology the present is the key to the past, in proposal economics it is the key to the future.

Logic

Careful logic is also a necessary ingredient in successful proposals because, once again, it helps allay the fears the audience has about the future. Since common forms of logic are widely accepted and used by educated people, their use enables the writer to lead the reader comfortably into a familiar frame of mind (pathos). Thereby logic projects a sense of the established past into the nebulous future.

The clear necessity of persuasiveness in proposal writing lends a special credence to the application of classical rhetoric in proposals, perhaps above all other technical genres.

As conceived by the ancients, particularly Aristotle, rhetoric is most useful in arguments that do not admit strictly scientific proof. Any projection into the future requires such an argument. But the more effectively the proposal imitates scientific proof by building a base of empirical and logical information, the more effective it will be in the hands of technical evaluators.

That the proposal should use logically precise arguments is perhaps a given, but what many writers overlook is the ways in which logic establishes the writer's trustworthiness (ethos) and influences the reader's emotions (pathos) [32].[9]

Graphics

Graphical presentation can enhance relevance by helping to accomplish the three goals of proposal writing identified by Herman Holtz [28], a leading authority on proposals to the government to:

* Create interest and desire
* Dispel the reader's fears about the uncertain future
* Demonstrate clear logic

Interest and desire are certainly attainable through visual means, especially through high-impact iconic depiction, such as can be achieved through color illustration and photography. As the graphic specialists Turnbull and Baird note, mass media experts consider the photograph to be the "best means for attracting a reader's eye to a page. . . . The emotions or reactions that are aroused as we view life about us can be aroused and catered to by photographs better than [by] any other means" [33, pp. 98-100].

In addition to stimulating interest, photographs help to achieve that second goal of promotional documents—counteracting uncertainties about the future by building a sense of a specific present reality. More often than not, however, proposals deal with products and tasks that are still in the design stage. Since it is impossible to photograph products that do not yet exist, proposers may develop and photograph realistic mock-ups or prototypes. Or they must find other means to harness the full power of graphics.

In engineering projects, schematics have the necessary simulating power: "They show on paper the circuit essentials, as if the circuit [itself] were available" [34, p. 10]. Time charts and organization charts likewise add concreteness in management plans [29, pp. 110-111, 133]. And, for non-technical audiences, a similar vividness arises from an "application illustration," a drawing or painting whose purpose is to demonstrate "the actual use of a product or equipment under conditions in which it was designed to operate, rather than to depict the exact details of the mechanism"; in a military proposal, for example, a jet fighter may be depicted in flight or a tank shown in jungle combat [35, p. 760]. Illustrations may thus "dramatize, as well as to communicate more effectively" [29, p. 246]. And, as we indicated in our discussion of the report, dramatization essentially means an enhancement of the reader's sense of rhetorical time in the present.

In using logic, especially as a means of bolstering the appeals of ethos and pathos, graphics can make a strong contribution to a proposal. Charts that show trends or changes over time are particularly effective because they take factual information from the past and present and project it into the future according to established logical norms. Tables can often be converted to line graphs that dramatize trends more effectively.

Both convention and the need for relevance, then, argue for careful and frequent use of graphical display in proposals.

THE LIMITS OF PERSONALIST RHETORIC

In all the genres of technical communication, the rhetorical conditions that arise from the demand for relevance suggest the need for dramatic presentations that lead to judgments and eventually to actions. To demonstrate the power of these demands, we have, in this chapter, drawn upon traditional rhetoric to restore the human presence that may have been displaced in our semiotic modelling in earlier chapters.

This theoretical position appeals to common sense by reminding us that technical communication is, after all, a human affair. It is about the relations between people, people who can be identified, analyzed, and written to. This personalist rhetoric stresses authors' ability to control the genres in which they write, to mold their texts to accommodate the needs of their audiences.

However, discourse theorists have in recent years grown wary of traditional rhetoric's insistence on the personalist approach to the relationship of author and audience, with its insistence on the intentional power of the writer. Semioticists, especially in Europe, have questioned many of the notions about writing and selfhood implicit in the rhetorical principles cited in this chapter.

The shift has been away from a personalized view of the text and toward a textualized view of the person. This view suggests that genres control writers and readers rather than vice-versa. In this way, roles are placed beyond the control of the producers and consumers of texts.

The meaning of this movement and its special significance for the practice of technical communication form the topic of the next chapter, in which we turn our attention from audience to author.

NOTES

[1] It also guides writers and students in the field. One day, for example, one of us overheard a group of technical writing graduate students discussing an audience analysis assignment, which required them to write up the same information for three different audiences. One student said that she was going to write first for an audience of high school students and then for an audience of professors. She spontaneously converted her thinking about audience into thinking about genres by saying, "I'll need a lot more information before I can turn the textbook chapter into a technical scientific paper."

[2] Pearsall's scheme derives ultimately from Aristotle [13], one of the first major theorists to suggest that public language is directed not so much to a specific figure as to a general type of person.

[3] The advantage of the active over the passive is one of the major truisms of a cognitive view of technical style. A cultural critique would show that, on many occasions, the passive voice and impersonal style are associated with a writer's professional identity and, if eliminated, would advance pathos at the expense of ethos. This is especially true of engineering and scientific writers. We discuss this matter at length in Chapter 7.

[4] No one in the community of technical communication scholars has pursued the purposeful question "So what?" more fervently than Donald Cunningham, who, in his various writings and in personal conversations and workshops, has contributed much to our understanding of purpose and audience in technical writing. See, for example, his book on manual writing [19].

[5] Classical rhetoric offers appropriate generic models for such structuring. Like technical reports, forensic discourse, as conceived by Aristotle, deals with the past as it affects the present and future. It involves, above all, establishing the truth about facts from the past, and thus it is the tradition that forms the basis for contemporary legal arguments. Did or did not Zeno murder a man? Did or did not the Memphis office violate company procedure? Like the technical report, the interest is in verification and confirmation. But that is not all. Aristotle [13, Book 1] also insists that the speaker must establish the just or unjust. Whether or not Zeno killed is less important than establishing if Zeno was justified in killing. Speakers concerned with past actions evaluate those actions and recommend that the audience accept the evaluation.

[6] Indeed, Aristotle shows that ethos operates through logos [13, Book 2].

[7] To some extent, the process of contextualization as here defined is a matter of tracing for the benefit of the audience the kinds of relationships graphed in our semiotic diagrams. These models, much like the x-y comparisons praised by Bertin, have *heuristic power*, the power to help the writer to discover potential subject matter and structures for evaluation and action.

[8]Linda Chavarria argues the negative for the principle, "Readers often interpret objective, colorless tone as coming from a writer who cares little about them or the subject" [27, p. 28].
[9]Forbes Hill claims that "Aristotle's approach was that ethos and pathos operate through logos" [39, p. 47].

REFERENCES

1. T. E. Pearsall, *Audience Analysis for Technical Writing*, Glencoe, Beverly Hills, 1969.
2. M. Keene and M. Barnes-Ostrander, Audience Analysis and Adaptation, *Research in Technical Communication: A Bibliographic Sourcebook*, M. Moran and D. Journet (eds.), Greenwood, Westport, Connecticut, pp. 163-191, 1985.
3. S. P. Sanders, How Can Technical Writing Be Persuasive?, *Solving Problems in Technical Writing*, L. Beene and P. White (eds.), Oxford University Press, New York, pp. 55-78, 1988.
4. R. E. Masse and M. D. Benz, Technical Communication and Rhetoric, *Technical and Business Communication: Bibliographic Essays for Teachers and Corporate Trainers*, C. H. Sides (ed.), National Council of Teachers of English, Urbana, Illinois, pp. 5-38, 1989.
5. J. Allen, Breaking with a Tradition: New Directions in Audience Analysis, *Technical Writing: Theory and Practice*, B. E. Fearing and W. K. Sparrow (eds.), Modern Language Association, New York, pp. 53-71, 1989.
6. T. T. Barker (ed.), *Perspectives on Software Documentation: Inquiries and Innovations*, Baywood Publishing Company, Amityville, New York, 1990.
7. C. Matalene (ed.), *Worlds of Writing: Teaching and Learning in Discourse Communities of Work*, Random House, New York, 1989.
8. J. C. Redish and D. A. Schell, Writing and Testing Instructions for Usability, *Technical Writing: Theory and Practice*, B. E. Fearing and W. K. Sparrow (eds.), Modern Language Association, New York, pp. 63-71, 1989.
9. R. Long, Writer-Audience Relationships: Analysis or Invention? *College Composition and Communication, 31*, pp. 221-226, 1980.
10. D. B. Park, Analyzing Audiences, *College Composition and Communication, 37*, pp. 478-488, 1986.
11. D. N. Dobrin, *Writing and Technique*, National Council of Teachers of English, Urbana, Illinois, 1989.
12. K. W. Houp and T. E. Pearsall, *Reporting Technical Information*, Macmillan, New York, 1968.
13. Aristotle, *Rhetoric*, in *The Complete Works of Aristotle: The Revised Oxford Translation*, 2 vols., J. Barnes (ed.), Princeton University Press, Princeton, 1984.
14. R. Hodge and G. Kress, *Social Semiotics*, Cornell University Press, Ithaca, New York, 1988.
15. J. Barzun and H. F. Graff, *The Modern Researcher*, Harcourt Brace, New York, 1970.
16. W. H. Beale, *A Pragmatic Theory of Rhetoric*, Southern Illinois University Press, Carbondale, 1987.
17. F. D'Angelo, *Process and Thought in Composition*, (3rd Edition), Little, Brown, Boston, 1985.

18. T. E. Pearsall, Panel Presentation on Technical Communication, Rocky Mountain Modern Language Association, El Paso, 1984.
19. D. Cunningham and G. Cohen, *Creating Technical Manuals*, McGraw-Hill, New York, 1984.
20. J. Bertin, *Graphics and Graphic Information Processing*, W. Berg and P. Scott (trans.), Walter de Gruyter, New York, 1981.
21. M. J. Killingsworth and K. Eiland, Managing the Production of Technical Manuals: Recent Trends, *IEEE Transactions on Professional Communication, 29*, pp. 23-26, 1986.
22. S. L. Fowler and D. Roeger, Programmer and Writer Collaboration: Making User Manuals that Work, *IEEE Transactions on Professional Communication, 29*, pp. 21-25, 1986.
23. J. Chew, J. Jandel, and A. Martinich, The New Communicator: Engineering the Human Factor, *International Technical Communication Conference Proceedings, 33*, pp. 422-423, 1986.
24. M. Dean, Make Your Reader the Hero of Your How-To Story, *International Technical Communication Conference Proceedings, 30*, pp. WE-3-6, 1983.
25. J. Maynard, A User Driven Approach to Better User Manuals, *IEEE Transactions on Professional Communication, 25*, pp. 16-19, 1982.
26. W. D. Skees, *Writing Handbook for Computer Professionals*, Lifetime Learning Publications, Belmont, California, 1982.
27. L. S. Chavarria, Improving the Friendliness of Technical Manuals, *International Technical Communication Conference Proceedings, 29*, pp. W26-28, 1982.
28. H. Holtz, *Government Contracts: Proposalmanship and Winning Strategies*, Plenum, New York, 1979.
29. H. Holtz and T. Schmidt, *The Winning Proposal—How to Write It*, McGraw-Hill, New York, 1981.
30. F. J. Evans, Jr., More Effective Engineering Proposals, *IRE Transactions on Engineering Writing and Speech, 3*, pp. 54-58, 1960.
31. R. J. Murdich, How to Evaluate Engineering Proposals, *Machine Design, 33*, pp. 116-120, 1961.
32. F. I. Hill, The *Rhetoric* of Aristotle, *A Synoptic History of Classical Rhetoric*, J. J. Murphy (ed.), Hermagoras Press, Davis, California, pp. 19-76, 1983.
33. A. T. Turnbull and R. Baird, *The Graphics of Communication*, (3rd Edition), Holt, Rinehart, and Winston, New York, 1975.
34. P. M. Beatts, Use Your Reader's Eyes, *IRE Transactions on Engineering Writing and Speech, 2*, pp. 6-11, 1959.
35. G. Magnan, Industrial Illustrating and Layout, *Handbook of Technical Writing Practices*, S. Jordon, J. Kleinman, and H. Shimberg (eds.), Wiley and Sons, Vol. 2, pp. 735-800, 1971.

CHAPTER 6

Generic Authors in Technical Communication

THE AUTHOR OF TECHNICAL COMMUNICATION

For years the emphasis in technical writing scholarship and textbooks fell upon the need for audience analysis. Recently, however, scholars have begun to consider more closely the perspective of the author, especially as the writer's role relates to his or her perception of the analyzed audience. The trend in the literature has been to recommend a personal, conversational "voice," a helpful "presence" that quietly and unobtrusively, yet steadily and surely, guides the busy reader past obstacles and frustrations toward the efficient accomplishment of technological actions.

This view of authorial presence shares much with our semiotic model. Above all, it treats the self of the author as a rhetorical entity, a projected sign that can be represented and interpreted. Generally, though, the literature does not go far enough in this direction, leaving the impression that the author of technical documents has a responsibility to linger personally within the text, much as a personal essayist would [1, 2].[1]

For students and novice writers of technical manuals, reports, and proposals, such an approach can be misleading. They may well wonder how the authorial voice of technical communication can vary from genre to genre and context to context even while the person writing remains the same. Confusion may develop over whether *ethos* (character in the ethical sense) or *persona* (an author's varying roles) controls the document's tone.

In this chapter, we develop a simple grammar-based semiotic to suggest how the best writers cultivate a genre-specific and context-conscious model of distance and involvement in the text and its preferred actions. As a point of departure, we consider the theoretical roots of current thinking on authorial rhetoric.

109

Ethos and Persona

Considerable literature has developed on the question of how the author's perspective should be treated in the style and rhetoric of a given document. In dealing with the problem, most scholars draw upon either the concept of *ethos,* derived from classical rhetoric, or the concept of *persona,* derived from the New Criticism, the major force in Anglo-American literary theory from 1940 to 1970. Both ethos and persona refer to the primary character of a text both in an ethical and in a technical sense, though ethos differs from persona theory in the way it treats the relation of these ethical considerations to the writer's personal values.

Ethos implies that the ethics of a text, which are made explicit in the arguments and judgments presented, are identical with the values of the author, the author's moral character. The technical character of the text, according to this perspective, is simply the "I," the author's sign, the means by which the values are set forth.

Persona theory, on the other hand, suggests that the narrator of a text is a character just like the characters about whom the narrator "speaks." The act of speaking demonstrates a set of values that are quite possibly distinct from those of the author, who maintains an ethical position outside the text.

To apply the concept of ethos to an analysis of technical communication would suggest a type of author ethically implicated by the text and morally committed to the agenda described therein. To apply the concept of persona to the analysis would imply an author who remains distant from the text and its characteristic audience, subject matter, and contexts.

Articles on technical writing have followed one or more of three general theoretical trends. They have:

1. Confounded the two concepts of ethos and persona [3].[2]
2. Recommended the adoption of personae, but have not been clear about the ethical responsibilities of the author or the general relation of the writer to the text [4].[3]
3. Recommended an aggressively personal approach to ethos without being clear about the technical means or the possible outcomes of such an approach [5, 6].[4]

What these writers agree upon is the need for a character to whom readers can relate, on whom they feel they can depend to fulfill their informational needs. No one, not even the personalists, suggests that technical writing is or should be autobiographical or that the reader's need for information presented in an easy-to-access form be sacrificed to the author's need for "self-expression." The "user-friendly persona" is clearly seen as a purposeful fiction projected by technical writers not for their own amusement but to help the audience read more effectively, and even the more assertive advocates of a personalized ethos have their limits (though these are not always clearly stated).

Beyond Ethos and Persona

Though the concept of persona that the recent literature borrows from the New Criticism departs from Romantic personalism [7], it nevertheless retains a vestigial humanism, a link with ethos. It retains, that is, a view of the text as a disembodied person. Structuralist and poststructuralist critics—Roland Barthes [8], for example, with his proclamation of the "death of the author" and Jacques Derrida [9] with his infamous critique of the "metaphysics of presence"—have done much to change that view.

Many technical documents, particularly technical manuals, are notoriously anonymous, so that the poststructuralist argument about the submergence of the author into a network of "intertextuality" is perhaps a better image for technical communication than it is even for the texts that the poststructuralists most frequently choose to analyze—literary texts brandishing the names of famous authors.

The existence of *intertextuality*—the dependence of a text on other texts with which it shares information, linguistic features, the marks of social conventions, and other codes [10]—is baldly obvious in technical writing. As David Dobrin notes in his well known chapter "Is Technical Writing Particularly Objective?", the first person singular pronoun is avoided in many company documents to increase their "fungability," the condition that allows portions of the text to be used in multiple documents by multiple authors in a clear instance of intertextual manipulation [11]. This year's proposal is next year's annual report. The actors may change, but the company and its actions endure.

How does the need to project an authorial presence vary in different documents and to what extent is the presence *conventional*—that is, *textual*—as opposed to being individually or personally motivated? A recognition of conventional authorial stances and an awareness of intertextuality should be vital components in the application of a model of technical communication to the problem of personality in technical documents.

Microstyle and Macrostyle

When students of technical writing are asked about the most confusing stylistic questions they face as writers, they often point to the use of *I* and *you*. Rather than dismissing this as a superficial concern of grammar, we can think of it as a lever to open a discussion on the shifting nature of authorial projection.

In the rest of this chapter, we will show how this simple point of grammar can become a conceptual tool to understand the conventions of authorial presence in technical writing.

Grammatical terminology can help the discussion at two levels, which we call *microstyle* and *macrostyle*. In the past, discussions of technical style too often got bogged down at the microstylistic level, with worries about specific problems of diction and sentence structure, such as use of nominalizations and the passive voice (see Chapter 7).

The recent literature on ethos and the persona, however, mounts the discussion at the macrostylistic level. This literature attempts to show how the use of the personal pronouns is related to other rhetorical techniques, such as humor, metaphor, and similar avenues to "user-friendliness" that involve the larger conception of authorial voice.

Our contention is that the two levels of style are intimately connected and that we can use the microstylistic concerns that often dominate the thinking of confused students (as well as the old approaches to teaching and editing technical writing) as a way of dealing with matters of macrostyle. The presence of the "I" in a discourse, for example, suggests the array of potential techniques that conventionally imply a strong authorial presence. Taking a hint from the rhetorical theorist Richard Lanham [12], who uses grammatical terminology to suggest the stylistic differences between the two major forms of public discourse—the "noun style" and the "verb style"—we seek to develop a "grammar of person" for technical writing, a way of showing that microstylistic concerns reflect larger issues [13].[5]

We begin by showing that the grammar of person is a convenient way to demonstrate the relations among the three major genres that we have modelled semiotically. Next, we will argue that the clues provided in our grammatical taxonomy suggest major differences in the degree of authorial presence appropriate in each of the genres and their attendant subgenres. These differences place in question many recent attempts to generalize about the "personalization" of technical documents. We are particularly concerned about the manual as a genre, for two reasons:

1. The manual has received a great deal of attention in the recent efforts to establish the need for personae in technical writing.
2. The manual is the most unique of the genres, in both microstyle and macrostyle.

We will return to the problems of authorial projection in the manual after having treated the three major genres in succession. The thesis of the final section on the manual is that the trend toward recommending the use of a strong persona may lead technical writers in a direction that hurts the overall effectiveness of instructional documents.

THE GENRES OF TECHNICAL WRITING: A GRAMMATICAL CLASSIFICATION

In the generic differences among the various kinds of technical writing lie important clues about an author's relation to an audience. A genre, in the simplest sense of the word, is a *kind* of writing. But, in a more complex sense, a genre is a set of styles, formats, and rhetorical approaches that are governed by strong conventions of practice.

Genre and Social Conventions

As Carolyn Miller points out, "genre study is valuable not because it might permit the creation of some kind of taxonomy, but because it emphasizes some social and historical aspects of rhetoric that other perspectives do not. . . . A rhetorically sound definition of genre must not be centered on the substance or form of the discourse but on the action that it is used to accomplish" [14, p. 151]. Miller also recognizes in her argument the insight of the literary critic Northrop Frye: "The study of genres has to be founded on the study of conventions" [15, p. 243].

Social relations gel into conventional relationships that come to be represented in the genres of discourse. We have shown in Chapter 4 that each genre involves a different relation of author to audience and a different outcome as a technological action. These relations are largely constituted by exchanges of power: in the proposal, the author is asking to be empowered to fulfill his or her own needs or the needs of the audience; in the manual, the author is giving knowledge that will empower the audience to use a product or system; in the report, the author is telling how he or she performed an action and placing the responsibility for further action upon the audience.

As authors shift from genre to genre, therefore, their attitudes toward the audience will shift as an effect of these changing power relations. Genre theory, therefore, offers one approach to the problem of how to locate the conventions of authorial projection in technical writing.

A Grammar-Base Classification of Genres

In Chapter 5, we showed that a simple rhetorical taxonomy can be based on the time conveyed in each of the major genres of technical writing: the report deals with past events, the manual with events presently unfolding, and the proposal with future events.

Translating *time* into *tense,* we can read this scheme as a *grammar* of technical communication. On an analogy with transformational grammar, each genre is reduced to a kernel sentence that we can later "transform" to allow for the various subgenres and alterations in tone. Table 4 summarizes the rudiments of such a grammar.

Though our grammar reduces the genres almost to the vanishing point, it allows us to provide a simple account of how the various kinds of technical writing are alike and different. The kernels share a number of grammatical qualities that suggest their overall companionship in the practice of technical communication.

They all use *do* as their verb. This indicates the common ground of action, as does the fact that all the verbs are in the active voice, though we will see that,

Table 4. Kernel Sentences for the
Genres of Technical Communication

Major Genre	Kernel Sentence
Proposal	*I will do this.*
Manual	*(You) Do this.*
Report	*I did this.*

especially in the report, passive transformations are possible, likely, and maybe even preferable. The tense of the verb changes from kernel to kernel, thereby suggesting the time conveyed in the different genres.

The kernels also all contain personal pronouns and the demonstrative pronoun *this*. We could have used lettered variables in these positions:

X will do y.
Do y.
X did y.

But we chose to use the pronouns both to avoid mathematical pretension and to take advantage of the linguistic notion of *shifters*. Shifters are words or transformations of words that point beyond the immediate linguistic context of the sentence either to something that has preceded the sentence in the discourse itself—the antecedent of a pronoun, for example—or to something outside of the discourse [16, pp. 2-7].

The pronouns *I* and *you*, which figure prominently in the discussion at hand, are always shifters that point to the context outside of the discourse. Unlike third person or demonstrative pronouns, which may point to nouns within the document (*he* may refer back to *John Smith*), *I* and *you* always point to the external context [17].[6] In written language, such words make rather strong demands on readers, who must grasp the exact relationship between themselves and the *I* who addresses them as *you* [18].[7]

Like *you* and *I*, the demonstrative pronoun *this*, which inexperienced or poor writers often use as a global reference to practically anything that has preceded it, is also a shifter. So, indeed, are changes in verb tense [19]. Shifts in the tenses of verbs indicate a change in the temporal context of the event in question. *I ran* situates the action in the past; *I am running* shifts into the immediate present.

The presence of multiple shifters in our simple grammatical scheme suggests the need in technical writing to make clear how an action depends on a variety of contexts—physical, psychological, social, and historical—which must be carefully specified in each document. By allowing slight changes in the relations of these shifters, we can suggest rather deep differences among genres and subgenres as we transform the basic kernels of our grammar.

Table 5 gives some sample transformations of each kernel. It will serve as a guide for our discussion of each of the genres.

The Proposal

The chief features of the kernel representing the proposal, *I will do this,* are the following:

1. the future tense of the verb,
2. the demonstrative pronoun as direct object,
3. the first person pronoun as subject.

Table 5. Transformations of Kernel Sentences to Show Subgenres

Kernel	Sample Transformations, Additions	Conditions
Proposal:	*We will do this.*	corporate identity
I will do this.	*May I do this?*	abject need
	This will be done.	effacement of proposer
	I will do this for you.	solicited proposal
	I will do this as you would do it.	unsolicited proposal
Manual:	*Our product does this.*	product-oriented
Do this.	*Our product allows you to do this.*	corporate identity, prominent author and user
	I will help you to do this.	prominent author and user
	This manual helps you to do this.	prominent text and user
	Do it this way.	policy/procedures
Report:	*I did this as I said I would.*	progress report/ final project report
I did this.	*This was done.*	effacement of agents, emphasis on actions and results (as in research reports)
	This can be done.	feasibility study
	You should do this.	recommendation report

Although the first and second of these features will change semantically—that is, the meaning of *do* and *this* will be specified in different ways—they will not change grammatically. And the grammatical alteration of the third feature is a slight, but rhetorically very important one. *We* may be substituted for *I* as an indication of multiple authorship or corporate authorship or as a rhetorical appeal for a sharing of values and actions with the audience. The clause, *as we all know,* for example, suggests a common ground of knowledge.

One of the scientists interviewed in a study of ethos by Preston Lynn Waller [20] uses *we* in the subject position in every non-passive sentence involving scientific action. Depending on the context, his usage shifts between significations of different versions of the corporate identity. At times, he seems to be using the term somewhat in the tradition of the editorial *we* (he calls it the "royal we")—a polite convention to avoid the first-person singular while suggesting it very strongly. At other points, he uses it to indicate participation of his graduate students and colleagues—*we at the Geoscience Department at my college.* At still another, he uses the term to conjure up a unity of scientific aspiration: "We will gain," he might write, "a new understanding of the geophysical phenomena under investigation." While retaining its grammatical consistency, therefore, the first-person plural subject is capable of a broad semantic and rhetorical variation.

The kernel sentence as a whole may be transformed to indicate additional differences in rhetoric. *May I do this?* would suggest that, unlike the scientific proposer, who is reviewed by his peers, the proposer is writing from a position of rather extreme powerlessness. All proposers who are requesting funds are in such a position to some extent, so that our kernel sentence always implies, *I will do this if I can get the money.* But this version of the sentence suggests something like an abject need. We can imagine that it would be an appropriate stance for a charitable organization.

We cannot, on the other hand, imagine a case in which the passive transformation of the sentence—*This will be done*—would be appropriate, for this sentence suggests a total effacement of the subject. And who would want to pour money into a project whose primary agent is absent from the proposal?

To distinguish between the major subgenres of the proposal, we need not transform the kernel grammatically but merely add some qualifying phrase or clause. The solicited proposal, one which responds to a definite Request for Proposal, may be represented thus: *I will do this for you.* The simple addition of the phrase *for you* suggests a corresponding change in the rhetoric—a very important change, for now the presence of the audience is directly invoked. Though the main clause retains the author-centeredness of the kernel, the revision suggests that the author is bound not only by his or her own needs or by the values of a research community but also by the needs of the reader as explicitly stated in the RFP. To use the term on which Herman Holtz and Terry Schmidt base their rhetoric for contract proposals, the document must be "responsive" [21].

The goal of such a proposal is to work toward what Kenneth Burke would call a "consubstantial relation" [22], in which the needs of the two parties are identified in a plan that would fulfill both. In grammatical terms, the *I* and *you* should be forged into a *we* by the proposal. This often comes about by a simple exchange of money for goods or services. *I* am a computer software engineer who needs money. *You* are an executive in a lucrative business that needs software. *We* can work together.

The unsolicited proposal, written to organizations like the National Science Foundation or the National Endowment for the Humanities and usually reviewed by peers of the author, may be represented thus: *I will do this as you would do it if you had the interest, ability, and/or inclination.* The *you* in the added clause is not so strongly invoked as is the *you* in the phrase added for the solicited proposal. This unsolicited *you*, as the additional clause suggests, is involved only in the review of the proposal, not in the project itself. The proposer must nevertheless seek a common ground of knowledge and values with this evaluator. The geophysicist mentioned above accomplishes this goal by developing a rhetoric of unity among research scientists.

No doubt some kinds of proposals fall into the gaps left by our classification of subgenres or represent a kind of combination that would require a specific analysis of the author's relation to the audience. For example, an internal proposal for resources and permission written by a staff member of a corporation and addressed to a manager might draw on any one of the three kernels already discussed—*May I do this?*, *I will do this for you*, or *I will do this as you would do it.* The choice of approaches would depend upon the kind of power exchanges and relationships of dependency that characterize the particular company in question.

The Manual

This genre should represent the most economical of technical documents and the purest embodiment of the principle of action-oriented discourse. Its purpose is to get a job done by providing information to a sometimes unwilling and even hostile audience. The introduction of unnecessary multiple shifters into such a text will add significantly to the cognitive strain of an already strained reader. In other words, overcomplicating the writing is going to make the user angrier than ever.

The grammar we recommend as a model for the manual genre is therefore the simplest and the least open to transformation. In fact, even the transformations that we have derived would be unacceptable to many manual readers, though, as we will show, they represent approaches that are quite common.

The first represents what was once the norm in the days of product-oriented documentation. *Our product does this* converts our original kernel *Do this* from the imperative to the indicative mood. It turns a command into a description, freezing the action implicit in the original kernel. Moreover, unlike the imperative, the transformed sentence fails to address the reader directly. A new emphasis on the product and on the corporate identity implied in the pronoun *our* is thus purchased at a considerable price—the loss of an emphasis on tasks to be accomplished and on the user who will accomplish them.

A second transformation—*Our product allows you to do this*—retains the emphasis on the corporate identity and the product while restoring the prominence of the user, but still the sense of action is diminished by the use of the indicative mood and the transformation of the original imperative command into an infinitive phrase—*to do this*—which functions as a noun.

A third transformation—*I will help you to do this*—brings the author or persona into a new prominence. We will have more to say about this approach a little later.

A fourth transformation—*This manual helps you to do this*—is similar to the third, but replaces the author's persona with a personified text in the subject position. Both the third and the fourth sentences suggest a guiding presence to help the user, and many commentators on the practice of manual writing indicate the need for such a rhetoric [23], even though it may slow or even obstruct the action.

For now, we may defer the dilemma by suggesting that the second, third, and fourth transformations, all of which convert the original imperative into an infinitive, represent an approach that is best suited for prefaces and introductions in manuals, while the body of their texts preserves the action-oriented elegance of *Do this*. In one of the transformations—*I will help you do this*—we have used the future tense to suggest further the appropriateness of sentences and rhetoric built on this model for introductions. *You will do this* might be an even more powerfully generative sentence in such situations.

Finally, in order to overcome an impression that the *Do this* is sacrosanct, we offer one of many possible transformations that could be made to suggest the various subgenres of the manual. *Do it this way* represents the approach of the policy and procedures manual.

The Report

Of all the genres of technical writing, the report is most variable in its grammar. As a genre, it enjoys a wide range that includes documents with subtle variations in the relationship of author to audience:

- The *data report* gives information with little or no commentary, usually in response to an audience who has contracted the authors to gather or produce the data experimentally.
- The *scientific paper* interprets data in support of a theory with which the audience of peers is thoroughly familiar.
- The *feasibility study* offers several options for action to an audience of decision-makers.
- The *recommendation report* provides an expert opinion on a course of action.

We could no doubt come up with other subgenres for the report. The common ground for all of them would be that they all include a presentation of information from the past relevant to decisions about future actions.

Each of the transformations in Table 5 represents a different subgenre of the report, within which other transformations and additions are possible. For the sake of space, we've included only a few of the possible transformations of the report kernel, to which we might add that all of the shifts from *I* to *we* in the section on proposals (and all the rhetorical consequences of that shift) apply just as well to the report.

In examining the sentences, we see a highly variable subject, which shifts from the agent of the action, the *I* or *we* of the progress report, to the natural object of the action, the passive subject *this* of the research report and feasibility study, to the *potential* agent of the action, the *you* of the recommendation report. The latter suggests the relationship of the recommendation report to the manual. It is also related to the proposal, as the future orientation of the verb *should do* indicates. The same may be said of the *can be done* in the kernel for the feasibility study, though the semantic difference indicates the difference in relation of author to audience in the two subgenres. The impartial author of the feasibility study indicates a range of possible options for action; the expert author of the recommendation report encourages particular actions. If this difference is pursued, the kernel for the recommendation report might be revised to read *I recommend that you do this* or *We recommend that you do this*.

What is important in understanding the range of the report is that the genre as a whole involves two crucial rhetorical steps:

1. the transition of action from past to future, and
2. the transfer of responsibility for action from the author to the audience.

The formulators of the data report present their information so that those who have purchased the information can interpret it. The authors of a scientific paper present their findings to an audience of peers who will use the information in further research. The authors of feasibility studies and recommendation reports provide options that the audience will judge and use as guidelines for various kinds of action.

The shifts of authorial presence involve *the degree of identification between author and audience,* which is the topic of the next section.

EVOCATION AND INVOCATION

The differences in perspective suggested by our analysis of the report author become even clearer if we place our grammar of person into a theoretical framework for understanding the way personality emerges in literary discourse.

Evocation and Invocation in Literary Theory

"Literature," argues the rhetorical theorist Walter J. Ong, "exists in a context of one presence calling to another" [24, p. 59]. In this communicative context, there is a kind of equilibrium. As authorial detachment increases, audience involvement increases in a compensatory response, and vice-versa: As the author's involvement increases, the audience becomes more detached.[8]

It is as if the text is a kind of vacuum longing for a human presence. As the *I* withdraws, the *you* fills it: but, if the *I* is fully present, the *you* remains distant.

The example Ong uses is the work of Poe, who could "never achieve so great a detachment [as that of Joyce or Faulkner], because his personal problems and neuroses show through" his various personae [24, p.59]. Ong uses the term *persona* almost in its literal meaning—*a mask*. The reader, confronting the persona, sees through the mask to the moral character (ethos) of Poe himself. Now in a position of moral superiority, the reader remains aloof and judgmental, detached and distanced from what is present in the story. With a text by Joyce or Faulkner, on the other hand, the reader must become creatively active, must essentially take the part of the author in order to "complete" the text or even to understand the action.

The text that is dominated by the *I* of the author we might call *invocative*, since it merely calls on or calls to (invokes) a reader. Poe's text is thus invocative. It puts the *I* on display for the reader to pass judgment.

The *you*-dominated text we might call *evocative*, since it calls forth (evokes) the reader and invites or even requires participation. Faulkner's novels are evocative.

Along the author-audience axis, then, we can imagine a continuum, on which lie various degrees of author and/or audience involvement, and which could be roughly represented by the diagram in Figure 32.

Invocation and Evocation in Technical Writing

In technical writing, the proposal stands at the invocative pole since its very purpose is to put the *I* on display as the subject of a proposed action that a relatively passive *you*—the proposal evaluator—agrees to fund or permit: *I will do this.*

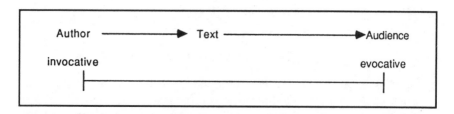

Figure 32. Author-audience continuum from invocative to evocative strategies.

The geophysicist interviewed in Waller's study [20], though he solves a number of grammatical and rhetorical problems by the consistent use of the *we* throughout his proposals, nevertheless runs the risk of de-centering his own person, whose credibility as the principal investigator on the project he is proposing is a crucial topic in any evaluation. He cannot rely totally on a corporate identity or an identification with the research community to provide this credibility. Another scientist whose proposals Waller studied uses a variety of personal pronouns in his work—*I, we, you*—but he occasionally faces a grammatical problem inherent in the use of shifters—ambiguous or remote reference. Clearly students and practitioners of technical writing face a rhetorical and grammatical difficulty here whose boundaries we have just begun to locate.

The transformations of or additions to the kernel also suggest that, as a genre, the proposal does not rest comfortably at the invocative pole. The reader is evoked in varying degrees depending upon the purpose and origin of the proposal. In the unsolicited proposal, the scientific author appeals to a community of scholars—*I will do this as you would do it*—and thus moves along the continuum toward the evocative pole. The solicited proposal, with its kernel *I will do this for you,* increases the degree of evocativeness.

The manual stands at the evocative pole, though, like the proposal, it cannot rest peacefully in its polar position. The author's presence may be asserted in varying degrees, as the transformation *I will help you do this* suggests.

Thus, in the revised diagram of Figure 33, we place the proposal and manual at separate poles along the continuum, but we must include as a qualification directional arrows that indicate the movement of each genre in the direction of the other.

We may now also add the report to the diagram (Figure 34). In its various subgenres and purposes, it partakes of both the invocative quality of the proposal—as in the various reports on a project which in fact may have been funded by a proposal effort—and the evocative quality of the manual, as in the recommendation report with its empowering of the *you.*

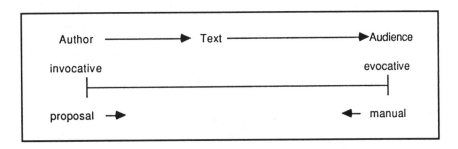

Figure 33. Relative position of the proposal and manual on
the author-audience continuum.

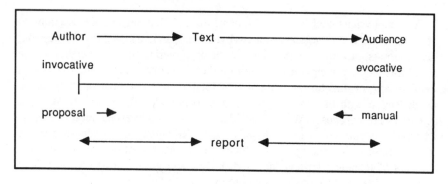

Figure 34. Relative position of the report on the
author-audience continuum.

The Manual and the Ultimate Weakness of Persona Theory

The range of the report may lead to questions about the applicability of the Ongian scheme. With a little imagination, we can still find a place along the continuum for most of the genre's various manifestations. But consider the kernel *This was done,* which seems to efface both author and reader, leaving a textual vacuum, or text without a human presence.

The existence of such a possibility, a vacant text, hints at a problem that comes to the fore in a closer analysis of the manual grammar. A reconsideration of the kernel *Do this* and its transformations points up the ultimate inadequacy of the Ongian equilibrium. In addition, it puts in question many of the assertions in the recent literature about the effectiveness of dominant personae in computer manuals (especially those of Bradford [4] and Rubens [6]).

In one sense, the presence of both *I* and *you* is obliterated in the manual. The kernel *Do this* suggests that the author has, in effect, disappeared and that the user, the "understood subject" of the imperative, has fused with the action, has become pure action or function.

Thus, in a sentence like the following, the imperative mood allows for the rhetorical absence of first- and second-person pronouns and suggests the illusion of a nearly pure action, the presence of a teacher and an agent being merely "understood": "Plug the other end of the cable into connector B." Nothing is gained here by saying "I recommend that you plug the cable in" or even "you should plug the cable in."

As a young freelance manual writer told Waller: "All my manuals are kind of anonymous. There are no attitudes in there, just procedures to follow." But her manuals, which rarely carry her name on the title page (they are literally as well as conceptually anonymous) and which give simple directions with a minimum of authorial intrusion, almost always receive high ratings on usability tests.

Even in introductions, where in most manuals the author greets the reader and gives a preview of coming attractions, the author's presence does not have to be prominent. Recalling Roland Barthes' proclamation of the "death of the author" and the ascendancy of the text, we might say that the text itself substitutes for the presence of the authorial voice. The manual writer interviewed by Waller says that, in her view, "the manual is this person" who is "sitting across the desk" from the user; "the author never comes in" [20].

Consider a sentence like the following from a manual written (anonymously) by the technical writer Joe Chew when he worked as a documentation specialist at Hayes Microcomputer: "Chapter Two, the Interbridge Manager Reference, gives a feature-by-feature description of the Manager" [25, p. viii]. This approach, involving a personification of the manual text, is modeled on the transformation *This manual helps you to do this,* though even the *you* disappears. Again, what would be gained by writing, "In Chapter Two, I give you a feature-by-feature description of the Manager"?

Authors of manuals, especially in the computer field, are experimenting widely with perspective and approach. They are asserting themselves, as Bradford notes, with the use of personal pronouns, with humor, irony, personification, analogy, metaphor, and quirky graphics [4]. The internationally known computer scientist Donald Knuth begins the preface to his innovative manual, *The TeXbook* [26] by invoking the "gentle reader." He refers to himself as "the author"—a device that, if anything, draws more attention to himself than would the simple use of *I*. He asserts quite openly that, while most manuals "make dull reading," his manual contains jokes to make the task more enjoyable. He warns that these "jokes can only be appreciated properly if you understand a technical point that is being made," so he admonishes the reader to "read carefully" [26, pp. v-vii].

In telling us that "people change as they grow accustomed to any powerful tool" [26, p. vii], Knuth is recognizing that the manual is a means of empowering the reader, of making the reader, in the words of Morris Dean of IBM, "the hero of the how-to story" [27]. And, in confronting us with a strong persona with his educational ideals and high expectations, Knuth offers a formidable implicit statement on the projection of personality in manual writing.

But some readers resist being cast in the role of hero and just want to get the job done. In another of Waller's interviews, Joe Chew warned that hero-making prose can seem "condescending." A manual writer, he says, should try to "project the attitude of being a teacher of intelligent students." Unfortunately, many manuals, in his view, "treat the user as an idiot; they go for the cutesy turn of phrase, thinking that this somehow increases the friendliness of the manual. I find this sort of thing offensive."

Consider, for example, this sentence from the *Nota Bene Tutorial* (1986): "For many of you, the purchase of Nota Bene marks your introduction to the world of computers and word processing, and perhaps the prospect of learning it all seems a bit daunting" [28, p. 3]. Intended to be comforting for new readers, this sentence is likely either to seem condescending or to increase nervous readers' apprehension by calling undue attention to their feelings.[9] The point is that authors stray from the operationalist kernel of the manual, the *Do this,* at great risk.

Both Knuth with the pun on "textbook" in the title of his manual and Chew in his conception of the role of the manual writer take the classroom teacher as a model. But our grammar suggests that the manual writer rarely needs to develop the formidable presence of the professor before a class. The legendary ego of academics may in fact harm the efforts of the manual writer who serves a difficult master—the easily frustrated user.

Technical writers frequently comment on the need to rid oneself of ego, a professional requirement that Jonathan Price [29] hints at in his inclusion of the *Diamond Sutra* in a list of suggested readings for computer documenters. In that sacred text, the Buddha teaches that no one who has advanced very far on the path to enlightenment "cherishes the idea of an ego-entity, a personality, a being, or a separated individuality" [30, p. 26].

From Macrostyle to Microstyle

In the next chapter, we will extend our discussion about the identity of the technical writer by taking up the theme of style. A frequent analogy for an author's style is *voice.* Finding one's voice, it is often said, is a key to effective writing.

What is the significance of style in securing a professional identity for those who serve technology through the written word? To answer this question, we turn from macrostylistics—the study of genres based on sentence patterns—to microstylistic analysis, the study of sentence patterns themselves, one of the oldest forms of study in the field of technical writing. In our attempt to broaden the study of technical communication, we are led into the field of discourse theory that overlaps with theories of human society.

NOTES

[1]On the relation of the personal essay to scientific and technical writing, see two articles by Killingsworth—"Science and Technology for a General Audience: The Personal Essay" [1] and "The Essay and the Report: Expository Poles in Technical Writing" [2].

[2]In "The Role of Ethos in the Theory of Technical Writing," Stoddard observes that "when technical writers and technical writing students perform their essential audience analyses, they must do so in relationship to themselves as writers, to the personae they adapt in any given communication act" [3, p. 232]. Persona theory would suggest, however, that "the personae they adapt in any given communication" could well be different from "themselves as writers."

[3]In a frequently quoted article, Bradford has noted, and approved, the trend toward the use of "personae" in microcomputer documentation. Technical writers' projection of

character, he asserts, "is as crucial to their success as is the accuracy of their information" [4, p. 68].

[4] Whitburn argues that, since science has come to be highly regarded in our culture, "a general depersonalization has swept through our entire society" [5, p. 302]. Communication has suffered and now looks to humanism for new life. Thus scientific and technical writing, that great outpost of impersonal objectivity, "might well be enhanced by a subtle infusion of personality . . ." [5, p. 306]. Why "subtle," for what ends, and in which genres? Rubens is more explicit in her claim that "impersonal, objective prose" is no longer a goal for *writers of computer documentation*; like Bradford, she suggests that they and their audience have insisted on "a more personal, 'user-friendly' style of scientific and technical writing" [6, p. 56]. But she does not carefully weigh the potential effects of this "user-friendliness."

[5] On the advantage of using grammatical terminology as a heuristic device in the theory and analysis of technical communication, see Killingsworth, "How to Talk about Professional Communication" [13].

[6] The linguist Emile Benveniste [17] goes so far as to suggest that such words should not be put in the same class with other pronouns. To use the terms of structuralist semiotics, which divides the sign into two parts, a signified (the concept to which the sign refers) and a signifier (the sign vehicle, a word, for example)—pronouns in the first and second person are profoundly free signifiers; that is, they may apply to anyone.

[7] As a demonstration of this difficulty even in speech, the linguist Otto Jesperson [18] cites the difficulty that young children have in learning to use *I* and *you*. Every day they face a startling variety of *I*'s—mother, father, Uncle John, Mr. Rogers—and are then expected to apply the word to themselves as well!

[8] Against these claims of Ong's, the personalist tradition of rhetoric would argue that too much authorial detachment breeds apathy in the audience, leaving the reader withdrawn into a "who cares?" attitude.

[9] In fact, this passage was called to our attention by an irritated user who had read an early draft of this chapter. Thanks to Jeffrey Smitten.

REFERENCES

1. M. J. Killingsworth, Science and Technology for a General Audience: The Personal Essay, *IEEE Transactions in Professional Communication, 25,* pp. 186-188, 1982.
2. M. J. Killingsworth, The Essay and the Report: Expository Poles in Technical Writing, *Journal of Technical Writing and Communication, 15,* pp. 227-233, 1985.
3. E. W. Stoddard, The Role of Ethos in the Theory of Technical Writing, *The Technical Writing Teacher, 11,* pp. 229-241, 1984.
4. D. Bradford, The Persona in Microcomputer Documentation, *IEEE Transactions on Professional Communication, 27,* pp. 65-68, 1984.
5. M. Whitburn, Personality in Scientific and Technical Writing, *Journal of Technical Writing and Communication, 6,* pp. 299-306, 1976.
6. B. Rubens, Personality in Computer Documentation: A Preference Study, *IEEE Transactions on Professional Communication, 29,* pp. 56-60, 1986.
7. R. C. Elliott, *The Literary Persona,* University of Chicago Press, Chicago, 1982.
8. R. Barthes, *Image, Music, Text,* S. Heath (trans.), Hill and Wang, New York, 1977.

9. J. Derrida, *Of Grammatology*, G. Spivak (trans.), Johns Hopkins University Press, Baltimore, 1976.
10. J. E. Porter, Intertextuality and the Discourse Community, *Rhetoric Review, 5*, pp. 34-47, 1986.
11. D. N. Dobrin, *Writing and Technique*, National Council of Teachers of English, Urbana, Illinois, 1989.
12. R. Lanham, *Analyzing Prose*, Charles Scribner's Sons, New York, 1983.
13. M. J. Killingsworth, How to Talk about Professional Communication: Metalanguage and Heuristic Power, *Journal of Business and Technical Communication, 3*, pp. 117-125, 1989.
14. C. R. Miller, Genre as Social Action, *Quarterly Journal of Speech, 70*, pp. 151-167, 1984.
15. N. Frye, *Anatomy of Criticism*, Princeton University Press, Princeton, 1957.
16. E. Mertz, Beyond Symbolic Anthropology: Introducing Semiotic Mediation, *Semiotic Mediation: Sociological and Psychological Perspectives*, E. Mertz and R. J. Parmentier (eds.), Academic Press, Orlando, pp. 1-19, 1985.
17. E. Benveniste, *Problems in General Linguistics*, M. E. Meek (trans.), University of Miami Press, Coral Gables, Florida, 1971.
18. O. Jesperson, *Language: Its Nature, Development and Origin*, Norton, New York, 1964.
19. R. Jakobson, Shifters, Verbal Categories, and the Russian Verb, *Selected Writings of Roman Jakobson*, Mouton, The Hague, Vol. 2, pp. 130-147, 1971.
20. P. L. Waller, "The Role of Ethos in the Writing of Proposals and Manuals," Doctoral Dissertation, Texas Tech University, 1988.
21. H. Holtz and T. Schmidt, *The Winning Proposal—How to Write It*, McGraw-Hill, New York, 1981.
22. K. Burke, *A Rhetoric of Motives*, University of California Press, Berkeley, 1950.
23. W. D. Skees, *Writing Handbook for Computer Professionals*, Lifetime Learning Publications, Belmont, California, 1982.
24. W. J. Ong, Voice as a Summons for Belief, *The Barbarian Within*, Macmillan, New York, pp. 49-67, 1962.
25. [J. Chew,] *Interbridge System Guide*, Hayes, Norcross, Georgia, 1986.
26. D. E. Knuth, *The TeXbook*, Addison Wesley, Reading, Massachusetts, 1986.
27. M. Dean, Make Your Reader the Hero of Your How-To Story, *International Technical Communication Conference Proceedings, 30*, pp. WE-3-6, 1983.
28. Anon., *Nota Bene Tutorial for Version 2.0*, Equal Access Systems, 1986.
29. J. Price, *How to Write a Computer Manual*, Benjamin/Cummings, Menlo Park, California, 1984.
30. Anon., *Diamond Sutra*, A. F. Price and W. Mou-Lam (trans.), Shambhala, Boulder, Colorado, 1969.

PART III
Communities

CHAPTER 7

Style and Human Action in Technical Communication

THE HUMAN FACE OF TECHNOLOGY

In our model of technical communication and in recent literature in the field, several concepts have evolved that suggest an emphasis on human action:

- Technical communication involves semiotic rather than dynamic action, interpretive human action rather than brute motion (Chapter 2).
- Media are varied according to cultural as well as cognitive needs (Chapter 3).
- The various genres of technical communication are social constructs ("codes") for transactions of knowledge and power (Chapter 4).
- Relevance—an awareness of a readership's present needs—is a leading indicator in the selection, organization, and presentation of content in technical communication (Chapter 5).
- In many genres of technical writing, the author actively creates the impression of a human presence (either that of the author or the ideal audience) within the text (Chapter 6).

These concepts all indicate a movement away from the old emphasis on technical writing as primarily "thingish" or oriented toward a description of a physical world and a movement toward an emphasis on technical writing as an effort to accommodate technology to the needs of human action [1-4]. Technical writing is, according to David Dobrin, *writing that accommodates technology to the user* [5, p. 54].

Style, Tone, and Attitude

In the current literature, the emphasis on technical communication as the human face of technology manifests itself primarily in discussions of *style*. In terms of technique, style is the set of principles by which words and sentences are chosen for a document. In terms of reader response, style is the voice of the text, that which makes it distinctive from other texts and which invites the audience to engage with it in some activity.

Like genre, style regulates the relationship between author and audience, for style is the primary means by which *tone*—the author's attitude toward the subject matter—is revealed. To create an attitude that encourages readers to act in a specific way requires a special style.

In technical communication, theorists, textbook writers, and working editors writing in the trade journals have raised an all but unanimous call for a style that reflects a world of people in action rather than a world of things in motion. Variously called "user-friendly," "active," "personal," "humanist," or just "plain," this style displays a preference for:

* active voice rather than passive voice sentences,
* mass-consumable language rather than discipline-specific jargon,
* direct address of the audience as "you,"
* personal rather than impersonal tone,
* an emphasis on narrative and the "verb style" and a sparing use of extended description and the "noun style."

In the development of this action-oriented style, technical writers assert themselves as advocates of the user of technology, a user interested in translating the text into specific technological actions.[1]

In this chapter, we review the primary characteristics of the action-oriented style, probe its historical development, suggest its limitations, and argue that it constitutes the master sign (or cluster of signs) associated with the emerging field of technical communication. Though the action-oriented style is not the only style used in technical discourse, and though it may not even be the best style on many occasions, it has become the style that technical writers claim as their own. It has thus become an identifying sign of membership in the community of technical communication—a point whose full implications we will explore in the last section of this chapter and in Chapter 8: A text is a model for action, and particular styles and text designs tend to support social structures that in turn support favored types of action in special contexts [6, 7].[2]

Thingishness: The Condition and the Antidote

The favored style of technical communicators tends to appear in the literature not as a *description* of practice but as an antidote for prevailing practice—a *prescription* for how technologists can improve their writing. The literature on style in technical communication thus takes on the aura of a campaign or a mission.

The condition for which the new style is prescribed has not been effectively named, though occasionally a term like "objective" is applied. Since objectivity is prized in technological circles, that is perhaps not the best term. Nor is "impersonal" a good choice, since it accounts only for a few of the symptoms of the offending style.

Instead, we use the term *thingishness*. It is comprehensive, easy to remember, and more suggestive than several alternatives in the rhetorical tradition. The condition it describes favors verbal structures that are often associated with objectivity but that in no way ensure it. These structures, which most technical writers and editors strive to eliminate, include the following:

- nominalization,
- strings of noun modifiers,
- passivity,
- indirectness,
- impersonality,
- unrelieved abstraction.

In their attack on these stylistic traits, technical communicators generally argue that they hamper the readability and usability of a document. Such a style is out of register with the world in which technical writing functions—a world of action, of connections, of applications, of relations, of people—all reduced by thingishness to a series of objects bearing questionable relations to one another and sharing no actions with each other. The potential harm implicit in thingishness may be revealed by applying the question of compensation: What is lost and what could possibly be gained by such a style?

Nominalization

In a thingish style, nouns replace other parts of speech, allowing things to supplant actions, judgments, and relations. In the terms developed by Richard Lanham [8], thingishness allows the "noun style," which is often associated with the sciences and social sciences, to replace the "verb style," which is preferred in the humanities and in journalism.

Nominalization describes a noun derived from a verb or an adjective [9, pp. 10-13]. The noun *operation*, for example, is derived from the verb *operate*, the noun *rationalization* is derived from the verb *rationalize* (derived in turn from the adjective *rational*). *Nominalization* is itself an example of the condition it describes, a noun made from a verb (*nominalize*) that is in turn made from an adjective (*nominal*).

Nominalizations are neither wrong nor bad in themselves; they allow stylistic variety when used in moderation. We can substitute the noun *need* for the verb *need*, *analysis* for *analyze*, *operation* for *operate*, and so on, to achieve the kind of variety that ultimately improves readability. Only when nominalizations stifle rather than enhance variety do they threaten to alter the way we represent the world in a technical document.

That world then becomes static, and the positions where readers could, as it were, insert themselves into actions described in the text are diminished in number and quality. Thingish prose strings together nominalizations with weak linking verbs and prepositions, as in the following example: "There is in our company a need for reassessment of the quantification practices that have become the first consideration of every experimental group since enactment of our new corporation management plan." Things predominate over people and actions, and a rather abstract form of description prevails over judgment and relevance.

Closely related to nominalization are verbs and adverbs like the notorious *prioritize* and the frequently used *operational*, which are formed from nouns and are used to replace more familiar action verbs and judgmental adjectives and subordinate clauses. These nominalizations in reverse add further to the feeling of a noun-dominated style and project the image of a thing-dominated world: "The aerojet project is operational under the guidelines prioritized in the institutional five-year plan."

Strings of Noun Modifiers

Noun-laden prose often eliminates even the vestiges of relationship by omitting prepositions. A "need for reassessment" becomes a "reassessment need." A "program for ensuring the rights for minorities" becomes a "minority rights insurance program."

Relations denoted by standard modification are obscured by this practice. In the last example, does "minority" modify "rights," "insurance," "program," or all three words that follow it? The same could be asked about any of the terms in the phrase except the last; "program" must, one assumes, receive modification from one, if not all, of the preceding words; they occupy adjectival positions even though they are nouns. Modification, however, does not flow smoothly through a group of words from first to second to third, so how can the reader discern the grammatical and semantic relation of an unbroken string of nouns?

Joseph Williams calls these constructions "compound noun phrases" [9, p. 26], but this term (in addition to being, like *nominalization*, an example of itself) suggests the presence of compounding conjunctions (like *and* or *but*) that, like prepositions, are unfortunately absent in a thingish style. Charles Stratton more accurately calls them "strings of noun modifiers" [10, pp. 449-450].

Both Williams and Stratton show how the judicious use of prepositions keeps the reader from imposing the wrong relations on these heaps of nouns ("the rights *of* minorities"). The use of adjectives can do the same and might also supply the judgment or the descriptive cues that readers need to discover the relevance of the material to their own needs ("a *democratic* program"). A modifying clause with a verb and object might show what actions connect all these things ("a democratic program *that ensures the rights of minorities* ").

No doubt words are saved in using strings of noun modifiers, but such brevity has its costs. The view of the world arising from sentences built around such constructions is a strangely unsettled image of things floating without relation and without effect upon one another.

Passivity and Impersonality

Most editors and teachers are aware that the overuse of the passive voice has been perceived as a major problem in technical writing [11].[3] We could argue that, because of its tendency to reduce the number of nouns in a sentence (through the omission of the *by* phrase in sentences like "The proposal was submitted by Professor Smith"), the passive voice might serve as an antidote to noun-heaviness. But this would be an evasion of the problem.

The thingish style is not just noun-heavy (as Lanham's use of the term "noun style" might unfortunately suggest), but specifically thing-heavy. And, as we all remember from grammar school, nouns represent persons and places as well as things. Persons and places disappear in a thingish style as part of the effort to obtain "objectivity," the effort to suggest that the events described could occur anywhere and could have been stimulated by anybody, that the results are "reproducible" (a quite appropriate message in scientific style but not in texts of technological action).

Like nominalization, the passive voice furthers this process by removing the agent from an active sentence [9, p. 13]. When we read, "The village was bombed," we wonder, "By whom?"—unless we are among the select few who can fill in the information without explicit cues (like the original readers of the Pentagon Papers, as the critic Richard Ohmann has shown in a memorable analysis of the passive voice in those documents [12]).

Things tend to be passive; they are objects; they are acted upon. Even such relatively active things as machines require human agency for their action. In documents related to laboratory research, animals (including human beings) become things, objects, receivers of actions: "The rat was injected with the virus"; "the dog's liver was extracted"; "the patient was prepared for surgery." The passive voice eliminates the active party and gives priority to the inactive.

In imitation of scientific style, people in business, industry, and government mimic this thingishness. Unlike the context for scientific writing, however, in which the subject position eliminated by the passive voice is always filled by more or less the same figure—the researcher—the context of technological action admits a variety of characters who could fill the subject position, including the researcher, the manufacturer, the user, and so on. If the text does not carefully specify the human agent of a particular action, the reader may not know who is responsible for carrying out that action.

In addition, the passive voice aids in the elimination of proper names, which are relegated to lists of references or documentary parentheses. First- and second-person pronouns vanish—which leads to the next characteristic.

Indefiniteness and Indirectness

Indefinite pronouns and nouns are common in the thingish style: *one, anyone, the individual.* The only personal pronoun that appears regularly is *it*, especially in its function as an expletive or meaningless word in what grammarians call indefinite or indirect sentence beginnings: "*It* is required of all personnel to report to the clinic"; "*it* was necessary that the mice be injected with the virus."

This convention aids nominalization and the passive voice in the attack on personality [1]. Additionally, when *it* is a true personal pronoun, it at least refers to a thing, but in indirect constructions it drifts toward no-thingness, a pure grammatical function, a superabstraction. Much the same is true of *there* when used in indirect sentence beginnings, such as "There is a need for more research." It forfeits its status as an adverb of place ("The book is over there") and becomes indefinite in the service of indirectness.

Such constructions violate the principles of the active style by leaving the reader disoriented without a foothold for action.

Unrelieved Abstraction

Thing is the most abstract word in English. What is a thing? Thingish writing indulges in flights of abstraction that rarely allow the reader's mind to descend to specific, material descriptions.

Consider this example from a work of literary criticism, a field that has been taken to task in recent years for adopting an impenetrable, pseudoscientific jargon:[4] "Only a radical transformation of bourgeois culture's 'base' can restore these 'superstructural' activities to their function as a communally self-referential labor quintessentially symbolizing man's humanization of nature."

To condemn this sentence for its "jargon" is a rather inadequate judgment because of the abstraction and range of possible meanings for the word *jargon* itself. To say "That's jargon" amounts to saying "That's bad." The real problem is that such passages sustain unrelieved abstraction at a very high level over several sentences or clauses.

The argument against such a style is this: A reader, especially a reader uninitiated in a particular discourse community, needs frequent, solid examples and other references to the facts of general life. These should mingle freely with abstractions; they should "break up" highly generalized passages in the same way that illustrations, graphic displays, headings, and white space break thick blocks of prose into manageable chunks. Specific examples permit readers to use their senses as well as their minds [13].[5]

Thingishness and the World of Technical Communication

From the perspective of a reader actively interested in translating the text into specific technological actions—the perspective to which the technical communicator strives above all to respond—thingish writing distorts the world in which technical communication fulfills its active purpose.

The qualities of the technical writer's world that thingishness weakly represents and that are represented more effectively in an action-oriented style include the following:

* a sense of action
* a sense of responsibility
* a sense of relatedness

A Sense of Action

Nominalization, passive voice, and other stylistic traits that increase the use of the verb *to be* and decrease the number and variety of action verbs suggest a static world instead of the world of technological action that gives rise to most technical communication.

Technical writing, as we have modelled it, represents and motivates action. Reports, in confirming and explaining events, suggest ways to plot future actions. Manuals help readers engage directly in activities. Proposals project actions into the future.

The overuse of *to be* freezes becoming into being. Actions, decisions, judgments, movements, changes, innovations, trends, all the motions and mobility of business, government, and industry become in the thingish style dead things to be picked over like so many bones beneath the archaeologist's shovel. No wonder that, in their recommendations for technical communication, the cognitive researchers Flower, Hayes, and Swarts advocate a "functional prose" that "is structured around a *human agent* performing *actions* in a particularized *situation*" [13, p. 42].

A Sense of Responsibility

Because passive, impersonal, indefinite, and nominalized constructions divorce agents from actions, a clear sense of who did what for which reason vanishes from the thingish document. This is a devastating distortion in a world where accountability is paramount—in the world of business or politics, for example.

It can also lead to serious imprecisions. When, for example, users of a computer software manual read, "The data are now entered," they must decide whether to perform the task themselves, to expect the computer to do it, or to believe that it has already been done.

Students writing research papers often report great difficulties in maintaining the conventional passivity and indirectness of academic prose while trying to give credit for certain findings to those responsible for them. How do writers distinguish between new findings offered in their reports and previous work reported in the literature? "It has been determined . . ." often prevails over "We have determined . . ." and "They have determined. . . ."

And how can readers sort out information about research currently underway and simultaneously carried out by different researchers? We might read, "The levels of acid rain are being measured in Canada, in New Mexico, and in Washington," and wonder, "By the same researcher? By three different groups?" The result is a confusion of responsibility.

As an example of the stylistic options posed by active and passive writing, consider the following rhetorical situation:

A team of researchers develops an hypothesis and does some preliminary experimentation that produces promising initial results. They want to create several distinct forms of writing that deal in different ways with their results. Call the results "this." It will be recorded in different styles in a scientific paper ("this was done"), a piece of science journalism written by a science writer ("the researchers did this"), a piece of science journalism written by one of the researchers ("I did this"), a manual for student assistants who will carry out similar experiments ("Do this"), and a proposal for further research using the same hypothesis and methods ("We should do this," *we* being the researchers and research assistants with overtones of cooperation from the funding agency and the general scientific community—see Chapter 6).

Figure 35 presents a diagram of the situation.

The different sentence forms shown as the representamen of the middle triad make different demands on the reader. "This was done" can only be effective if the conventions of the text imply clearly established roles for the author and audience. If both are scientific researchers, for example, as in the case of most journal articles, nothing much is lost by use of the passive construction. Both the writer and the reader just assume a standard agent (the researcher) for each action and focus on the action, not on the identity of the agent.

All of the other genres, however, require identification of the agent; and these are the kinds of writing most frequently initiated by technical writers.[6] Science journalism restores the narrative structure of human action to the scientific process—it makes a story of science—so that unfamiliar material may be

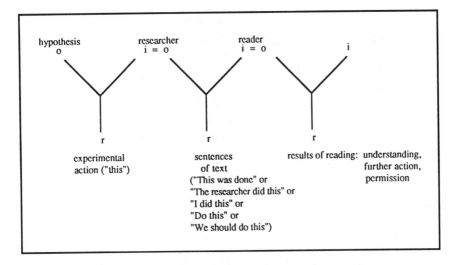

Figure 35. Semiotic diagram of stylistic options in
reporting a scientific result.

presented in a familiar form, thus enhancing the understanding of the reader. And, as we suggested in Chapter 6, technical genres that take action as their ultimate end, especially manuals and proposals, must situate users directly in the action.

When the passive sentence—"This was done"—predominates in action-oriented genres, readers have to insert additional links in the semiotic chain. They have to create sentences that translate agent-less action into motivated action to understand their own relation to what is supposed to be going on. As the reader's representative in this process, the technical writer provides the necessary links and thus speeds up the process of converting information to action.

A Sense of Relatedness

Technical writers synthesize information from various disciplines and often "translate" the language of one discourse community into language that a different readership can understand. They help to make possible the flow of information among the various components of the American technological system, the "military-industrial complex"—academic science and engineering, government, the military, business, and industry—components which, since World War II, have become ever more interdependent.

Among these groups of interests, technical communication facilitates *transactions*, "simultaneous and mutually dependent interaction[s] between multiple components," and thus conforms to the "systems view of life" that, according to the physicist and popular philosopher Fritjof Capra, prevails in general systems theory, the science of ecology, and relativistic physics [14, p. 267].

By omitting important connectives and terms of relationship—prepositions, subordinating conjunctions, participles, pronouns—the writer of thingish prose destroys the reader's sense of how things relate to one another. By overusing the verb *to be*, the author reduces to states of being actions that would have related subject and object. By depersonalizing prose, the writer eliminates cues about the relationships of specific people with things and events.

THE HISTORY AND IDEOLOGY OF
THE ACTIVE STYLE

The advantages of the action-oriented style, which we have just reviewed, are well documented in the textbooks and literature on technical communication. In addition to enhancing readability for a particular kind of reader, this style, in combination with the iconic-mosaic design of documents, has become the mark of a particular writing culture. It has come to reflect the professional identity of the technical communicator.

The study of style thus presents a clear opportunity for demonstrating the relationship between the theory of language (or signs) and the theory of culture (sociology or anthropology).

Ideology and Style

Building upon the work of Volosinov [15] and Bateson [16], the Australian semioticians Robert Hodge and Gunther Kress, in their book *Social Semiotics*, highlight the link between style and *ideology*, the system of ideas and myths upon which group identity is formed and maintained: "Just as individual acts of semiosis are organized by systems of signifiers of power and solidarity, so also are the relationships between groups in a broader social formation. These broader signifying systems . . . link the social organization of semiotic participants with social organization on a larger scale. Any group of any size needs markers of group membership to give it identity and cohesion, and to differentiate it from other groups" [17, p. 79].

Hodge and Kress go on to say that such groups are not marked by any single sign, but rather by a "cluster" of signs—a *style*—of which a good example is the cluster of textual traits outlined so far in this chapter. Whereas the traits of the thingish style have been traditionally associated with research science, social science, and engineering, the traits of the active style are associated with human services and with information-centered professions, including technical communication.

The active style favored by the leading practitioners of technical communication has been shaped by the social and historical development of technology. What social alliances do technical communicators seek in their use of the active style, and what social advantages can they hope to achieve for their readers, for their subject matter, and for themselves?

Against the tendency to pronounce the universal effectiveness of any particular cluster of stylistic options, we would do well to recall once again the principle of compensation: *In every transformation of a text, something is lost and something is gained.* A socially conscious version of the compensation theory might be rendered thus: *In adopting any stylistic option, an author appeals to some groups and offends others.*

The Plain Style Revolution

Every group, then, has a set of shared values and a collective image—an ideology. Group members use ideology to ease internal conflicts and take refuge from external strains.

As a process of building identifications through discourse practices [18], style generates ideology in emerging groups and maintains ideology in mature groups. Ideology is the historical group's self-image; style is the set of techniques used to achieve that image. Group identity depends for its success upon a distinctive style, whether a lifestyle or a written style [19, 20].[7]

When technical writers discuss the active or "plain" style, they commonly hint at their ideological interest by using political language. "Now the plain language movement is becoming a *revolution*," writes Alan Siegel, for example [21, p. 222; italics added]. The beneficiary of this revolutionary style is not the writer, according to Siegel, but the reader, the "consumer" [21, p. 223].

The use of the economic term *consumer* suggests that the plain style represents an effort to take technical discourse out of the academy and the laboratory and into the marketplace. It removes technology from an elitist environment and gives (or sells) it to the people. It thus requires that language should be "radically simplified," preferring words that are "familiar" rather than "unfamiliar," sentences that are "short" rather than "long," a format that is "unintimidating" rather than formidable, and a tone that is "personal" rather than "impersonal" [21, p. 223].[8]

Texts written in this style thus aim for a mass appeal that is all business, with personal pronouns and human subjects to stand for real actors with real responsibilities, active verbs to reflect an active life. Its open layout and graphical features guarantee that busy readers will not have to spend too much time reading before they put the information to work.

Like the iconic-mosaic design format in its affinities with advertising, the plain style has come to be associated with the broadest range of readers and consumers. It is the chosen medium of both public information systems and business communication. When technical writers advocate the use of plain English, they take on the role of "consumer advocates"; they implicitly commit themselves to an alliance with the community of business and human services; they become the reader's representative, interloping in what would otherwise be an elitist culture of contemplative science and high technology. Their ideology is marked by a concern with openness and freedom of exchange, key values that they share with those who profess "the business creed" of free enterprise [19].

Roots of the Revolution

Contrary to Alan Siegel's suggestion, however, the plain style revolution and the ideological connection between technical writing and the business culture are not new. Early in the history of technical communication, the rising class of business people and crafts workers exerted a strong stylistic influence on scientific rhetoric and practice.

The first scientific journal, the 17th-century *Proceedings of the Royal Society,* was edited not by a scientist, but by an entrepreneur [22]. Under his editorial direction, the authors of papers for the *Proceedings* resolved, in the words of the Royal Society's official historian, Thomas Sprat, "to reject all amplification, digressions, and swellings of style, to return back to the primitive purity and shortness, when men delivered so many *things* in an equal number of *words,"* to promote "a clear, naked, natural way of speaking, positive expressions, clear senses, a native easiness, bringing all things as near mathematical plainness [as possible], and *preferring the language of Artisans, Countrymen, and Merchants, before that of Wits and Scholars"* [quoted in 23, p. 352; italics added]. Thus, during the emergence of modern science in the 17th century, the Royal Society adopted the language of artisans, countrymen, and merchants in a show of solidarity with the burgeoning middle classes and the activities of trade and the crafts.

Our stereotype of science as the "leading edge" of cultural development in the West is, in many ways, a projection back onto history of the high status that science enjoys in our own time. In fact, science has followed as well as led. Historians and sociologists of knowledge have shown how advances in science have often depended upon technological innovation, especially developments of instrumentational techniques. The historian Derek de Solla Price argues that "historically the arrow of causality is largely from the technology to the science" [24, p. 240]. The anthropologist Clifford Geertz effectively summarizes the thrust of Price's research: "Science owes more to the steam engine than the steam engine owes to science; without the dyer's art there would be no chemistry; metallurgy is mining theorized" [25, p. 22].

Examples of scientific dependence on technology proliferate in history. The philosopher Don Ihde suggests that, through the use of improved lens technology, Galileo was able to construct a telescope that, though crude by later standards, was yet able to inspire "a new *technologically mediated* paradigm for scientific vision" [26, p. 55]. The crafts provided not only instrumentation to be used in scientific inquiry, but also information, as Thomas Kuhn indicates: "early fact-gathering is a far more random activity than the one that subsequent scientific development makes familiar. Furthermore, in the absence of a reason for seeking some particular form of more recondite information, early fact-gathering is usually restricted to the wealth of data that lie ready to hand. The resulting pool of facts contains those accessible to casual observation and experiment together with some of the more esoteric data retrievable from established crafts like medicine, calendar-making, and metallurgy. Because the crafts are one readily accessible source of facts that could not have been casually discovered, technology has often played a vital role in the emergence of new sciences" [27, pp. 15-16].

It is likely that, in an implicit acknowledgment of technology's power and influence, Sprat and his Royal Society colleagues theorized that the plain style created a flexible and widely acknowledged medium of exchange. Like money, the plain style enhanced transactions between science, technology, and business. Then, as now, the aim of technical communication could be classified as "transactional" [28].

In the 17th century, the problem was to take a first step toward opening science to interchange with business and the crafts by making scientific style simple and direct. In the 20th century, new demands have been made on technical style due to the influx of an ever-increasing and ever more varied population of nonspecialists into the audience of writings about technology, which has become, in turn, ever more rooted in scientific research.

The new audience is interested in technical information mainly as a means of improving actions. In creating an action-oriented style, technical writers have found it useful to restore an agent for that action, thus putting the reader into a framework of productivity, making the reader (in the words of Morris Dean) the "hero of the how-to story" [29]. The presence of the reader demands, in turn, a reliable guide, a "persona"—hence the renewed call for a personalist approach to authorship [30].

The Emergence of Elite Discourses in Science and Technology

Clearly, then, much has happened in the history of the science-technology-business relation in America between the 17th century and now. We can summarize developments this way: Just as science divided into basic science and applied science, technology has divided into technical engineering, technical salesmanship, and human services, with technical communication emerging as a recent division of the latter category.

The efforts of the Royal Society to create in the 17th century a style that could facilitate transactions among science, technology, and business was opposed by a contrary movement. By the late 18th century, with the fall of the feudal aristocracies, came the rise of a technological elite—the physical scientists and engineers who within a hundred years would control the tempo and the direction of the industrial revolution.

The formation of elite communities of knowledge was at least partly the result of the explosion of scientific activity and information that accompanied the birth of the new nation-states in the revolutions against monarchies. In the United States, "the number of persons who could be considered fully engaged in technical pursuits has been growing at extremely high annual rates, running at levels between two and three times greater than the secular growth rates in population," with the result that "every generation in America . . ., even that of the Founding Fathers, probably felt that it was living in a golden age of science and technology" [31, p. 34]. As Derek de Solla Price points out, at any point in history, Americans "could look back and say that most of the scientists that have ever been are alive now, and most of what is known has been determined within recent memory" [32, p. 14].

This phenomenal increase of people, activity, and knowledge in the technical professions has been accompanied by an extreme division of labor—a result that, since the time of Adam Smith, economists have come to expect in the social organization of capitalist/republican cultures. By the end of the 19th century and the beginning of the present century, at which time the total number of techno-logical specialists was "roughly doubling every decade" [31, p. 35], elite cadres of researchers had appeared in almost every scientific field—each group managing the production of knowledge in ever-narrowing specialties, the gaps between which were becoming ever more difficult to traverse.

In these elite groups, technical styles developed that separated the groups from each other and from business and industry.[9] As the 20th century unfolded, scientific discourse became even more intensely theoretical and thus ever more difficult for nonspecialists to comprehend or apply in practical affairs [22].

This development would have made relatively little difference in the evolution of science and technology had not changes in the relationship of science and industry come about in the organization of industrial society. Up until the 19th century, science and technology had worked side by side, influencing one another through the medium of advanced instrumentation. Designs for technical instruments were shared, for the most part, by amateur investigators laboring in communities of inquiry that were, at best, loosely coordinated by friendly correspondence, the publications of groups like the Royal Society, and books (which were published in short runs and were thus expensive and hard to come by).

In fact, it was not until about 1840 that the term "scientist" came to be used to distinguish "a new corporate consciousness amongst those who pursued science, either as a career or as a dominant and serious leisure activity" [33, p. 194]. By the end of the 19th century, for the first time in history, scientific investigation had become thoroughly professionalized. The process was complete when scientists all but universally established their credentials by graduate education and appointments to university faculties.

Professionalization of scientists made it possible for people to think of science as a unified, centralized, and organized body of knowledge. To make the findings of this newly organized research community useful in technological pursuits, and thereby to create a kind of mutually sustaining super-technology, the emerging system of production now required only the mediating discipline of engineering, which by the end of the 19th century had also been professionalized and was flourishing both in the academy and in industry.

The engineers who arose to fulfill the demands of a technological society in the 19th century, were, like scientists, committed most heavily to inquiry and the invention of new technological objects. However, their origins in the mechanical fields and the crafts (which, in the early days, included medical practice) assured that their style and their outlook were more tempered by human interest and entrepreneurial economics. They could thus create an effective link between scientific discovery (oriented to understanding the natural world) and product development in industry (oriented to usefulness in the human world) [34].[10]

The Alliance of Science, Business, and Government

Coeval with the professionalization of science and engineering, a new alliance had begun to form that would eventually prove irresistible even to academic scientists and would lead them to question their own insulation from practical, worldly matters. In the years just before the Civil War, the famous inventor Eli Whitney sold the rifles he designed and manufactured to the United States government. This kind of transaction normalized in the decades that followed as the nation became a military power and as the military bonded with science and private industry.

The connection of science, private enterprise, and government was a possibility that scientists recognized even in the age of Leonardo da Vinci and Galileo, but it had yet to be realized at the institutional level [26, 35, 36].[11] By the early 20th century, however, private industry itself was working to close the gap between scientific research and technological innovation, as the example of Thomas Edison and the founding of the General Electric Corporation demonstrates [34]. As the period of the world wars approached, one prominent scientist and research manager could exclaim, in quite appropriate metaphors, "The research men of a nation are not isolated individuals but an organized and cooperating army" [quoted in 34, p. 158].

But the military eventually found that it could not always wait for the chain of transactions among science, engineering, and industry to produce the information and products needed for modern technological war. The transfer of knowledge/ power was too slow and too poorly controlled to guarantee effective applications of state-of-the-art technology to military needs.

During World War I, the pace of technological development increased dramatically when, for the first time, the cooperation of academic scientists was enlisted by the public sector (the executive branch of the government) to provide the innovation and theoretical prowess needed to bolster the productivity of technological industry in the private sector [34]. The diagram in Figure 36 offers a schematic view of this alliance. After the first world war, the partnership between the academy and industry lagged. With the Great Depression, it all but dissipated entirely [34].

It was not until World War II, with its mass mobilization, that the full cooperation of academic scientists with the military and the major economic institutions of the capitalist economy in America—business, industry, and representative government—crystallized into a tight technological structure [37, pp. 405-406]. President Eisenhower would later call this alliance "the military-industrial complex," though, as a number of commentators have suggested, the former general was remiss in leaving the university and its professional scientists out of the equation [38]. The Manhattan Project and other such scientific/military ventures drew researchers into a new economic relationship with government, one that to this day dominates the production of knowledge in the United States [39]. The historian Jay M. Gould rightly argues that "World War II, and the development of the atom bomb in particular, effected a radical alteration in the relation of science to industry, because it dramatized the tremendous power that organized scientific research could bring to industry. The scientist of today has [thereby] achieved

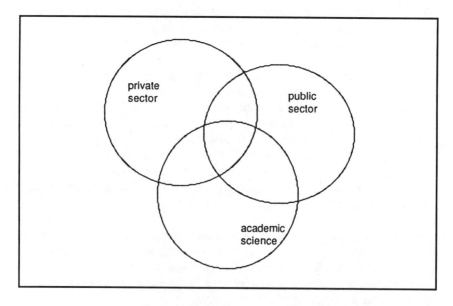

Figure 36. The alliance of science with the public and private sectors
in World War I and World War II.

much greater status. . . . He is courted and acclaimed and can command favored positions in business, government, and university circles" [31, p. 89].

Technical Communication and the Military-Industrial Complex

This crystallization of institutions in the 1940s and 1950s produced the needs that eventually resulted in the formal profession of technical communication. Manuals had to be written for poorly educated military personnel required to use sophisticated technology. Proposals had to be written so that academicians and business people could plug into the elaborate new systems of government funding. Reports were needed to link together the various cooperative institutions in a network that required accountability and that demanded submission to surveillance practices. Technical editors and writers worked to assure that the technological elite presented information in a style that promoted rather than hampered cooperation [40].[12]

Thus, the field of technical communication was born in the overlapping space shared by the major institutions of the military-industrial complex, as Figure 37 suggests.

In this effort to increase the speed and overall efficiency of communicative transactions in research and development, America has provided the geographical

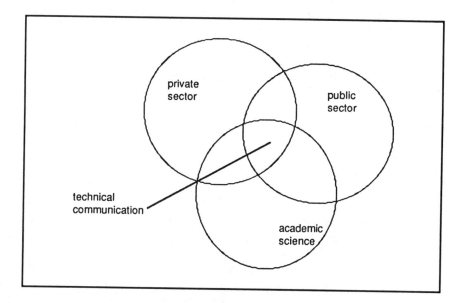

Figure 37. Place of technical communication in the "military-industrial complex."

and socio-economic setting for the renaissance of the 17th century attempt to create an effective transactional style that appeals mutually to the technical professions and the human services professions.

The Humanist Critique of Technical Style

The high status of engineering in the new technological complexes has encouraged another revival—the resurgence of the dream of early scientists to develop a mechanical measure of stylistic effectiveness. While Sprat and his colleagues longed for a neat fit between words and things and a "mathematical" plainness in style, modern language engineers have sought to mechanize the production of discourse with the tool of readability formulas.

Few scholars in technical communication have approved of this turn of events. For example, Michael Halloran and Merrill Whitburn argue that "recent thinking about plain language is rooted in the same simplistic positivism as 17th century views on the plain style and that this positivistic view of the universe is even less appropriate in our own time than it was in the 17th century" [30, p. 60].

According to Halloran and Whitburn, the Royal Society distorted the Ciceronian ideal of plainness—the use of unadorned, familiar language from the public realm to reach a broad public interested in informed action—by failing to allow for effective modulations among levels of style and by insisting on a simplistic correspondence between words and things (hence the charge of "positivism").[13] Likewise, modern advocates of formulas like Flesch's readability score and the Gunning Fog Index hope to develop a simple technology of style, a kind of labor-saving device, that will free the technical writer from the rigors of interpretation and rhetorical analysis by creating an absolute guide to clarity in language, a style that allows the reader to view the world through a lens (language) so transparent that it may be unnoticed.[14]

Against this trend, the humanists Halloran and Whitburn hope to restore a more complex Ciceronian view of style, concentrating in particular on rhetorical techniques for exhibiting "personality" in prose. They argue that the person—both reader and writer—has been abstracted out of the formulas for achieving clarity in writing. In phenomenological terms, we can say that the formulators of stylistic plainness seem to have forgotten that clarity must always be clarity *for someone.* Halloran and Whitburn rightly criticize the focus on the relation of words to things because this approach turns the writer's attention away from the need to communicate with the reader on a person-to-person basis. They argue for a re-personalized technical writing.

Ironically, in building their final argument not upon the authority of history and the Ciceronian tradition but upon personal experience, Halloran and Whitburn use an appeal that correlates strongly with the Royal Society's own original rationale for stylistic reform. "By the authority of . . . our own experience as consultants to industry," they write, "we reject impersonality as a stylistic ideal" [30, p. 67].

They claim that "the assumptions that led to both scientific method and the modern plain style have had a profoundly negative impact on communications today." "Revolutions," they continue—using the political term and thereby revealing their ideological slant—"are typically reactions against excesses, and the reactions are often as excessive as the original abuses" [30, p. 67].

The Humanist Influence in Technical Communication

As a reaction to attempts to "engineer" language and communication, then, a humanist role has emerged, predicated upon the need to put the personality back into technical prose. Technical writers, many of them trained in departments of English and other humanist disciplines, have entered the field primarily as *human interest* or *human factors* specialists, representatives of users and readers neglected in the engineer's privileging of the technological object.

Through the medium of technical communication, the humanities, with their interest in discourse and culture, have discovered a new social function in the alliance between the academy and the major sectors of production in the American economy. Their position is represented in Figure 38.

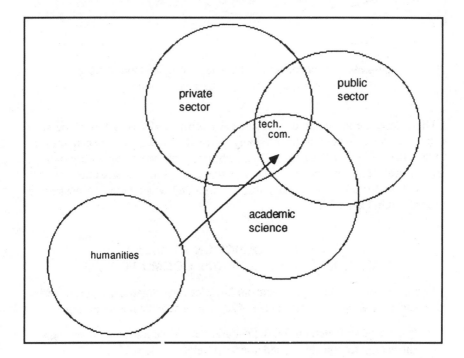

Figure 38. Potential avenue of influence of the humanities in the socio-economic complex of technological research and development.

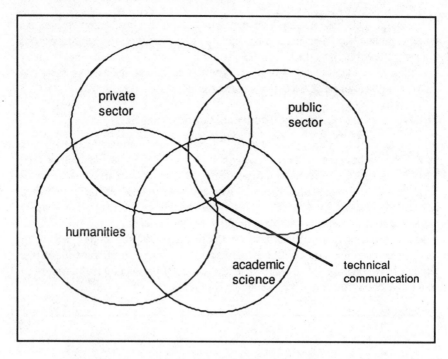

Figure 39. Full incorporation of the humanities into the matrix of
technological research and development.

Ultimately, the academic disciplines, once considered "ivory tower" pursuits
without a useful function in the economy—indeed a burden on the economy, the
"wits and scholars" whose style Thomas Sprat dismissed as unworthy of 17th
century science—may soon (for better or worse) be fully incorporated into the
matrix of cooperation in technological research and development, a possibility
represented in Figure 39.

LINKING PRODUCTION TO USE:
THE WORK OF THE TECHNICAL COMMUNICATOR

Our analysis of the active style and our historical investigations suggest a series
of conclusions about the development of style in technical communication:

1. The active style demonstrates a cultural preference for expressions connot-
   ing action, responsibility, and a sense of relatedness.
2. In adopting this style as a master sign of identity, technical communicators
   have historically demonstrated solidarity with the business and consumer
   side of the production/consumption relation in the political economy.

3. In industry and research—the heart of technological production—technical communicators thus tend to serve as representatives of funders, users, and consumers.

4. But a total identification with those represented is impossible, since the users themselves expect technical documents to represent expert technological production.

5. The great problem of technical communication is therefore to develop a style that fills the gap between technological production and consumption, simultaneously representing both sides (or any two or three sides) of a technological transaction.

6. As different groups move into positions of social ascendency—such as business and the crafts in the 17th century, or science in the 20th century—they will create new demands on technical styles.

7. Thus, stylistic determination in technical communication is never simply a matter of relating words to things, but is always embedded in sociopolitical processes—a matter of relating groups of people to other groups of people.

The Technical Writer's Dilemma

Relating groups to other groups is never a simple matter. The sensitive position of the technical writer in the world of high technology is captured effectively in Cunningham and Cohen's thoughtful book on technical manuals. A provocative cartoon depicting "the publication dilemma" shows the technical writer as a Janus-faced young man, looking simultaneously left and right. To one side stands the engineer (a bald-headed authoritarian), saying "I gave you all the info you need—tons of it. It's right there on your desk: my logs, the printouts, the specs, the lab reports, the drawings. . . ." To this aggressive and thinly veiled accusation of incompetence, the technical writer responds (somewhat limply): "I can't make heads or tails of it. . . ."

To the other side of the beleaguered writer stands the user (a young woman, blond with glasses removed and eyes staring forward with a dazed expression), with whom the writer carries on a frustrating dialogue:

Writer: "I don't know what you want. . . ."
User: "I don't know. Period!"
Writer: "I need help!"
User: "You think *you* need help!" [41, p.8]

The idea is that technical writers—and the same would be true of technical editors or graphic artists—must represent two constituencies. During the writing process, they must take the part of the reader in discussing form and content with the technical experts who have developed the product; and, in the final version of the document, they must represent the product to the reader.

In each case, they take the part of one person in the face of another, much as a lawyer first represents a client before the judge and then represents the judge to the client by explaining the court's decision. The attorney, like the technical writer, is above all a *meditator*.

If the division of labor in technological research and development has been organized as a chain, as we have suggested, then the technical communicator serves as the final link that connects the developed product to the activities of the user. Cunningham and Cohen's graphic depiction of this phenomenon of mediation and representation conforms nicely with another version of the link-making analogy—the Peircean concept of the sign, which, in the words of Joseph Ransdell [42], is a "Janus-faced entity": like the technical writer in the cartoon, the sign looks both ways; it represents one entity to another, but in various stages of the process it is "reversible."

History teaches that products and technology can influence the direction of research and development just as surely as research and development can influence the production and distribution of products. Likewise, the technical communicator, by taking the part of the end user, can influence the process of technological production.

In light of these relationships of mediation and representation, Figure 40 offers a diagram of the technical writer's position using the triadic chain that, in Part I, we derived from Peircean semiotics.

Why Are Technical Writers Needed?

Why is the technical writer needed in this semiotic exchange? Why is there a gap between the technical expert and the user of the manual that must be filled by

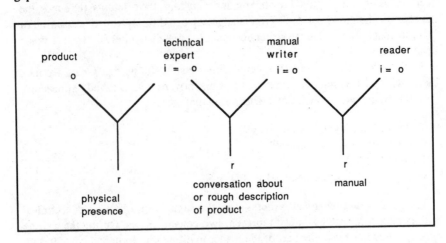

Figure 40. The position of the technical writer in
the semiosis of the manual.

this "middleman"? Is this simply another instance of "Parkinson's Law," the tendency of administrators to create more work for more administrators [39, p. 10]?

Many managers will say that the problem is simply that technical experts can't or don't have time to write; others will say that it is a matter of economics: It is cheaper to pay a writer to document a product than to pay an engineer or programmer to leave the work of product design and take up the work of writing.

But the problem goes deeper than this. The user wants the knowledge of the technical expert, but, because of the division of labor in our social organization, this user is not sufficiently trained to grasp that knowledge without the help of an "interpreter." Or, to use a different metaphor, we can think of Francis Bacon's famous equation of power and knowledge: the user wants the power of the technical expert, and the technical writer serves as a transformer of that knowledge/power, "stepping it down" to make it "safe" for ordinary use.

Technical experts often resist this use of their power, viewing this contact with the world of business and marketing as an intrusion upon their productivity. The technical writer is seen, at best, as the tool by which the technical expert makes connections with the world at large, thereby ensuring continued funding for technological projects. Very likely, the technical writer will feel resistance (or perhaps frustration) from both sides, the supply side (the technical expert) and the demand side (the user).

Technical communication, then, is deeply involved with the supply and demand relation of knowledge/power.[15] Sensitive politics govern that relation, which may in turn have a strong effect on the style of a technical document produced under competing institutional forces.

To insist upon the active style is to assume a particular kind of author—a helpful provider of information, a committed communicator. But, in the rhetorical situation we have diagrammed in Figure 40, the person most capable of giving reliable information is not the writer but the expert whom the writer represents.

The division of the communication situation in this manner creates a complication that reaches beyond the simple communicational model suggested in cognitivism with its two willing partners in the communicative act. Indeed, we might argue that the insertion of the writer in the exchange between the technical expert and the reader is an attempt to revive a communicative act that has already gone bad.

To get a better sense of how the preferred style of technical communication relates to the styles of other disciplines and communities, we will turn in Chapter 8 to an analysis of the different groups of writers and readers with whom technical communicators most frequently have contact. From our analysis of these "discourse communities" will emerge a picture of the overlapping subcultures that influence the production of texts in the technological world.

NOTES

[1] This is the perspective normally affiliated with what we called the "cognitive" approach in Chapter 3.

[2] To compare action in a text to action in a society, as we have been doing in our exploration of the analogy between texts and technology, is to create a simultaneous discourse on texts and societies; one term of the metaphor (society is a text) informs the other (a text is a society or a social model), and vice-versa, as Paul Ricoeur suggests: "In one way the notion of a text is a good *paradigm* for human action, in another the action is a good *referent* for a whole category of texts. With respect to the first point, human action is in many ways a quasi-text. It is exteriorized in a manner comparable to the fixation characteristic of writing. In becoming detached from its agent, the action acquires an autonomy similar to the semantic autonomy of a text; it leaves a trace, a mark. It is inscribed in the course of things and becomes an archive and document. Even more like a text, of which the meaning has been freed from the initial conditions of its production, human action has a stature that is not limited to its importance for the situation in which it initially occurs, but allows it to be reinscribed in new social contexts [as in the use of *precedents* in legal actions]. Finally, action, like a text, is an open work, addressed to an infinite series of possible 'readers.'" Ricoeur goes on to suggest that, in similar fashion, "certain texts—if not all texts—have as a referent action itself. In any case this is true of the story." He refers to Aristotle's position in the *Poetics:* "the *mythos* of tragedy, that is, at once the fable and the plot, is *mimesis,* the creative imitation of human action. Poetry, he also says, shows men in action. The analogy between text and action does not seem risky once we can show that at least one area of discourse has action as its subject; it refers to it, redescribes it, and repeats it" [6, pp. 160-161]. In our view, the genres of technical communication number among those texts that have action as a referent or subject matter. Technical reports, manuals, and proposals are *about* technological action. Moreover, they are themselves a form of technology; they are *instrumental.* This claim places us in the philosophical camp to which Richard Rorty [7] has attached the name "textualists," theorists whose exploration of the metaphor of action as text (and text as action) leads them toward the conclusion that the theory of the text illuminates the theory of action and vice-versa. As Rorty suggests, most textualists privilege the text in this theoretical transaction; that is, they tend to "read" society as a series of rhetorical interactions. Our position in this regard might be called "actionist" since we are concerned with how an understanding of the kinds of action typical of technology might influence the kinds of text characteristically produced in a technological society. But we would also want to maintain the textualist view of a dialectical relation of text and action in which one constantly influences the other. Another name for our position might be "textual realism."

[3] Thomas Warren compiled an annotated bibliography on the subject that was so long it had to be published in two installments in the *Journal of Technical Writing and Communication* [11].

[4] Though many such complaints are merely the work of writers who lash out because they don't understand current scholarship in the field, much of the dissatisfaction is justified. On the one hand, we find recent developments in the theory of literary criticism to be quite useful for the analysis of nonliterary as well as literary works. But, on the other hand, we believe that theorists are obligated to show how their work is useful in a language that is generally comprehensible.

[5]Or, to use a figure that is perhaps more cognitively suggestive, we can say that examples lead readers to a correct version of the sense experience and save them the trouble of having to do their own "translating" of abstract material into language more immediately accessible to the senses since such translation is a necessary part of meaning-making in the human mind [13].

[6]Technical editors often work with scientific authors, of course, and have to respect the right of these authors to their conventional use of the passive.

[7]There are two dominant understandings of ideology in social theory. One is the Marxist and neo-Marxian view of ideology as "false consciousness," a layer of myth that masks an essentially materialist and economic reality. The business creed of "free enterprise," according to this view, uses the democratic value of freedom to hide—even from those who benefit from the ideology—the domination of the working class by the capitalist bourgeoisie. A second understanding of ideology has developed in functionalist sociology, especially in the American school. This is the view we are explicating here: all groups have an ideology and use it as a means of attaining exclusivity and solidarity. We feel this is a view more analytically powerful than that of neo-Marxism, since, in claiming that ideology is both false and all-pervasive, analysts in the leftist tradition allow no room for self-critique. The functionalist view admits that ideology is all-pervasive, but it leaves open the possibility of critique from different perspectives, thus avoiding the need to condemn or privilege any single ideology absolutely and preserving the possibility of applying the compensatory principle, even to one's own viewpoint.

[8]The term *plain* suggests naturalness—a kind of appeal typical of ideologues: Our way is more natural than your way.

[9]It is a commonplace in ideologically-oriented stylistics to note that, at this point in the development of the science-business interchange, business communication began to show the effects of the newfound power and influence of science. Today, we find in business widespread use of jargon borrowed from the technical elite. Thingishness in style may, in this sense, be seen as a form of *scientism,* an effort on the part of outsiders to show the rhetorical signs of solidarity with the scientific community—that is, the "neutrality" or "objectivity" of the scientific outlook.

[10]The historian of technology David F. Noble writes: "The research activities of trade associations, semiprivate institutes, independent contractors, government bureaus, and private foundations provided an essential service and subsidy to the expanding science based industries. They were not, however, equipped to meet all of the scientific requirements of modern industry. Only the nation's colleges and universities, with their unequaled research facilities and trained personnel, were prepared for this. Given the proper organization and spirit of cooperation, the universities could provide an unlimited amount of applied research for industry. More important, as the traditional site of fundamental scientific research, they could support the basic investigations upon which industrial research was grounded, investigations which only a handful of the largest corporations could afford to underwrite. 'The very existence of the great research programs of industry,' William Wickenden maintained, 'is predicated upon the existence of a vast army of free, disinterested and even impractical researchers at work in the laboratories of colleges and universities.'" In addition, Noble points out that "the universities could do what no other research agencies, within or without industry, could do: they could reproduce themselves" through their teaching mission [34, p. 128].

[11]Derek de Solla Price has shown that Galileo first intended his telescope to be used for military, not astronomical, purposes. Don Ihde calls this a "striking glimpse of the early modern version of the science-industry-military combination" [26, p. 53]. Ihde also notes another such instance in the work of Leonardo, who "offered himself to a series of wealthy aristocrats and warring lords," presenting in one letter an early prototype of the technical proposal: "I know how to build very light, strong bridges, made to be easily transported so as to follow and at times escape from the enemy. . . . I know techniques useful in invading a territory, like how to drain water out of moats and how to make an infinite number of bridges and covered walkways. . . ."As Ihde notes, "This engineering science is as wedded to the 'military-industrial complex' as any Eisenhower ever dreamed of!" [26, pp. 194-195]. No wonder that Kelly MacIntire has described the artist that produced the Mona Lisa as an early technical writer [35]. As early as 1714, nation states were seeking help from science and technology. In that year, the English Parliament passed an act to fund a reward for anyone who developed a device for measuring longitude at sea [36, pp. 49-50].

[12]Essentially, it was the increased pace of the technological transaction that both linked science and engineering and caused the division of the field of engineering into technical experts and technical communicators. As Max Weber suggested in his famous analysis of bureaucratic organization, the chief advantage of productive systems that employ a radical division of labor is speed [40]. It is not surprising, then, that contemporary advocates of stylistic plainness insist on accommodating a busy readership.

[13]The whole question of the Royal Society, plain style, and Cicero has become entangled in misunderstanding. First, the Royal Society presumed, like Francis Bacon, that they were following Seneca, not Cicero, when they eschewed the rounded, periodic sentences, figurative language, and impressive vocabulary of the grand style then associated with Cicero. But this presumption sprang from a Renaissance misunderstanding of Cicero, who often praised lucidity and simple diction. He advocated the use of all three styles—the plain, the middle, and the grand—each in its proper place. He integrated them with the three functions of an oration—to teach or inform (logos), to conciliate (ethos), and to move (pathos) an audience. According to Cicero, the orator should conciliate in the beginning of an oration, which requires the middle style, so that the impression created is neither too simplistic nor too elevated during the first contact with the audience. It is best, next, to teach in the middle sections of an oration when arguments are advanced. This section requires the plain style, which makes it easy for the audience to follow the arguments. Only at the end of the oration does the inspired orator resort to the grand style, so that the audience is moved to accept the position that has been clearly explained and is now eloquently rendered. The grandness of the grand style should be relative to the other styles used in the oration and relative also to the occasion as well as the tastes of the orator's age.

[14]As Idhe argues, "Textual transparency is hermeneutic transparency, not perceptual transparency" [26, p. 82]. A lens is a technology that allows a user to perceive a world through a transparent medium; any adjustments that are necessary are handled by the user's body, as it were. The glass lens thus allows perceptual transparency. The old analogy between the lens and the text does not work, however, since the text requires the user's mind to make adjustments in order to envision (interpret) a world. If the text employs a style that is familiar to the user, many such adjustments may seem automatic, thus creating an illusion of perceptual transparency, which in fact corresponds to Ihde's "hermeneutic transparency," a condition arising from interpretive activity that is, for the user, unconscious, part of the user's background. Hermeneutic transparency depends upon the user

being so familiar with a particular line of argument that conscious interpretation is unnecessary. Like Ihde's concept, our semiotic model demonstrates the ultimate superficiality of demands for "clarity." A text is not a window, a simple visual technology, or even a lens. It is a complex chain or network of signs, every link in which requires both representation and interpretation. However, habitual use of clusters of signs does create, for members of a discourse community, a sense of relief from interpretive rigor. The demand for clarity is always politically loaded, since what is clear for one group will not necessarily be clear for another with a different background.

[15]The economist Fritz Machlup makes an interesting point about this kind of involvement: "As an economy develops and as society becomes more complex, efficient organization of production, trade, and government seems to require an increasing degree of division of labor between knowledge production and physical production." Machlup wants to suggest a further division, a new category of people involved in the production of knowledge—the *transmitters* of knowledge: Besides "the researchers, designers, and planners," those who produce new forms of information, we should also keep in mind "the executives, the secretaries, and all the 'transmitters' of knowledge in the economy" in any analysis of the production of knowledge, since knowledge is not merely information but is better defined as information used for a particular purpose and given a high value [39, pp. 6-7]. Certainly, technical communicators would be included in Machlup's "transmitter" category and would thus be included in his broadening of the concept of "production" to cover not only physical objects but also "knowledge."

REFERENCES

1. M. Whitburn, Personality in Scientific and Technical Writing, *Journal of Technical Writing and Communication, 6,* pp. 299-306, 1976.
2. C. R. Miller, A Humanistic Rationale for Technical Writing, *College English, 40,* pp. 610-617, 1979.
3. S. M. Halloran, Technical Writing and the Rhetoric of Science, *Journal of Technical Writing and Communication, 8,* pp. 77-88, 1979.
4. S. P. Sanders, How Can Technical Writing Be Persuasive?, *Solving Problems in Technical Writing,* L. Beene and P. White (eds.), Oxford University Press, New York, pp. 55-78, 1988.
5. D. N. Dobrin, *Writing and Technique,* National Council of Teachers of English, Urbana, Illinois, 1989.
6. P. Ricoeur, *The Philosophy of Paul Ricoeur: An Anthology of His Work,* Beacon, Boston, 1978.
7. R. Rorty, *The Consequences of Pragmatism,* University of Minnesota Press, Minneapolis, 1982.
8. R. Lanham, *Analyzing Prose,* Charles Scribner's Sons, New York, 1983.
9. J. Williams, *Style: Ten Lessons in Clarity and Grace,* (2nd Edition), Scott, Foresman, Glenview, Illinois, 1985.
10. C. Stratton, *Technical Writing: Process and Product,* Holt, Rinehart, and Winston, New York, 1984.
11. T. L. Warren, The Passive Voice Verb: An Annotated Bibliography, Parts I and II, *Journal of Technical Writing and Communication, 11,* pp. 271-286, 373-389, 1981.

12. R. Ohmann, *English in America: A Radical View of the Profession,* Oxford University Press, New York, 1976.
13. L. Flower, J. Hayes, and H. Swarts, Revising Functional Documents: The Scenerio Principle, *New Essays in Technical and Scientific Communication: Research, Theory, Practice,* P. V. Anderson, R. J. Brockmann, and C. M. Miller (eds.), Baywood, Amityville, New York, pp. 41-58, 1983.
14. F. Capra, *The Turning Point: Science, Society, and the Rising Culture,* Simon and Schuster, New York, 1982.
15. V. N. Volosinov, *Marxism and the Philosophy of Language,* L. Matejka and I. R. Titunik (trans.), Seminar Press, New York, 1973.
16. G. Bateson, *Steps to an Ecology of Mind,* Granada, London, 1973.
17. R. Hodge and G. Kress, *Social Semiotics,* Cornell University Press, Ithaca, New York, 1988.
18. K. Burke, *A Rhetoric of Motives,* University of California Press, Berkeley, 1950.
19. F. X. Sutton, S. E. Harris, C. Kaysen, and J. Tobin, *The Business Creed,* Harvard University Press, Cambridge, Massachusetts, 1956.
20. C. Geertz, *The Interpretation of Cultures,* Basic Books, New York, 1973.
21. A. Siegel, The Plain English Revolution, *Readings in Technical Writing,* D. C. Leonard and P. J. McGuire (eds.), Macmillan, New York, pp. 222-229, 1983.
22. C. Bazerman, *Shaping Written Knowledge: The Genre and Activity of the Experimental Article in Science,* University of Wisconsin Press, Madison, 1988.
23. M. Whitburn, et al., The Plain Style in Technical Writing, *Journal of Technical Writing and Communication, 8,* pp. 349-358, 1978.
24. D. J. de Solla Price, *Little Science, Big Science . . . and Beyond,* Columbia University Press, New York, 1986.
25. C. Geertz, *Local Knowledge: Further Essays in Interpretive Anthropology,* Basic Books, New York, 1983.
26. D. Ihde, *Technology and the Lifeworld: From Garden to World,* Indiana University Press, Bloomington, 1990.
27. T. S. Kuhn, *The Structure of Scientific Revolutions,* (2nd Edition), University of Chicago Press, Chicago, 1970.
28. T. E. Pearsall and D. Cunningham, *How to Write for the World of Work,* (3rd Edition), Holt, Rinehart and Winston, New York, 1984.
29. M. Dean, Make Your Reader the Hero of Your How-To Story, *International Technical Communication Conference Proceedings, 30,* pp. WE-3-6, 1983.
30. S. M. Halloran and M. D. Whitburn, Ciceronian Rhetoric and the Rise of Science: The Plain Style Reconsidered, *The Rhetorical Tradition and Modern Writing,* J. J. Murphy (ed.), Modern Language Association, New York, pp. 58-72, 1982.
31. J. M. Gould, *The Technical Elite,* Augustus M. Kelley, New York, 1966.
32. D. de Solla Price, *Little Science, Big Science,* Columbia University Press, New York, 1963.
33. C. A. Russell, *Science and Social Change in Britain and Europe, 1700-1900,* St. Martin's, New York, 1983.
34. D. F. Noble, *America by Design: Science, Technology, and the Rise of Corporate Capitalism,* Oxford University Press, New York, 1977.

35. K. MacIntire, "Leonardo DaVinci: Technical Communicator," M. A. Thesis, Memphis State University, 1989.
36. D. J. Boorstin, *The Discoverers*, Vintage Books, New York, 1983.
37. R. Weigley, *The American Way of War: A History of United States Military Strategy and Policy*, Indiana University Press, Bloomington, 1973.
38. M. Kenney, *Biotechnology: The University-Industrial Complex*, Yale University Press, New Haven, 1986.
39. F. Machlup, *The Production and Distribution of Knowledge in the United States*, Princeton University Press, Princeton, New Jersey, 1962.
40. M. Weber, *Economy and Society: An Outline of Interpretive Sociology*, University of California Press, Berkeley, 1978.
41. D. H. Cunningham and G. Cohen, *Creating Technical Manuals: A Step-By-Step Approach to Writing User-Friendly Instructions*, McGraw-Hill, New York, 1984.
42. J. Ransdell, Peircean Semiotic: The Grammar of Representation, unpublished manuscript, 1988.

CHAPTER 8

Communities of Discourse

THE CONCEPT OF DISCOURSE COMMUNITY

The actions that generate writing and the actions that result from reading are determined not only by the individual needs, preferences, and abilities of authors and audiences, but also by the social and historical conditions under which the writing and reading take place. Even the internal mental states of writers and readers are largely a product of interaction with external conditions.

Drawing upon these principles, the advocates of *discourse community theory* argue that no stage of the writing process is ever purely a lonely, individual pursuit, the work of a self, an atomistic *I* addressing a particular *you*. Writing and reading are activities motivated and guided by community perceptions and values.

Writers Always Collaborate

In a study of *invention*—the stage of the writing most likely to be connected with the origin of the text and thus most given to claims of individual originality—the rhetorical theorist Karen Burke LeFevre argues that, in one way or another, *writers always collaborate,* if not with friends and colleagues, then with an internalized "other," a mental representation of some absent mentor, former associate, or ideal reader [1]. In this expanded sense, invention is always a social process [2-4].[1]

If it is true that, in inventing and drafting a text, writers cooperate with others, it is likewise a sure bet that, in later stages, these writers are even more likely to collaborate. Their writing will be revised, edited, and packaged in processes that are often beyond their control. Even in cases where they maintain a high degree of individual control over the text (by means of desktop publishing, for example), their writing is derivative in the sense that it is based on reading and education, two of the major avenues of socialization in a literate culture.

Language itself is the social medium *par excellence*. It not only provides the means by which social interaction is carried beyond the physical level; it also shapes the individual mind in the forms of a culture, as social psychologists since George Herbert Mead have clearly recognized [5]. From its very first inkling in the writer's mind, then, writing is influenced and limited by the culture that the language represents and perpetuates.

Community as a Communication System

The text also comes under the influence of the subcultures, or *communities,* with which writers identify and for which they write. In the theory of discourse communities, old notions about the relationship of author to audience as *one mind seeking another* give way to a more integrative concept of the author and audience as *participants* in a *communication system.* In this network of interconnected and interrelated individuals, each participant is both a sender and a receiver of information, and each is already a complex of social formations.

This communication system goes by various names. Building upon the concept that linguists have called the "speech community" [6], the Frankfurt social philosopher Jürgen Habermas refers to "communication communities" in his influential theory of communicative action [7,. 8].[2] For reasons no doubt governed by his own communication community, the literary critic Stanley Fish prefers the name "interpretive communities" [9].

Theorists in the field of English composition use the term "discourse communities" [10, 11], which we think is best for our purposes. It is broad enough to include many forms of text generation (speaking, writing, and graphics) and text recognition (listening, recording, and interpreting), but it is narrow enough to exclude more basic forms of communication (electronic or animal signals, for example, and other instances of what we called "dynamic actions" or "dyadic signs" in Chapter 2).

The clearest definition of the concept as it is understood by composition theorists is given by James Porter: "A 'discourse community' is a group of individuals bound by a common interest who communicate through approved channels and whose discourse is regulated" [12, p. 38]. Porter lists a number of examples of "professional, public, or personal discourse communities": "the community of engineers whose research area is fluid mechanics; alumni of the University of Michigan; Magnavox employees; the members of the Porter family; and members of the Indiana Teachers of Writing" [12, p. 39].

He explains that each community uses certain "forums" as approved channels of discourse; examples include "professional publications like *Rhetoric Review, English Journal,* and *Creative Computing;* public media like *Newsweek* and *Runner's World;* professional conferences (the annual meeting of fluid power engineers, the 4Cs); company board meetings; family dinner tables; and the monthly meeting of the Indiana chapter of the Izaak Walton League" [12, p. 39]. Each forum is constrained by a set of "rules governing appropriateness to which members are obligated to adhere"; these rules may be "more or less apparent, more or less institutionalized, more or less specific to each community" [12, p. 39].

The interest in discourse communities has spawned an emergent school of theorists, the "social constructionists" [13], and has awakened many researchers in composition and professional writing to the possible relevance of methodologies pioneered by ethnographers [14, 15].

The Relevance of Discourse Community Theory to Technical Communication

For engineers and scientists, who take co-authorship for granted, and for technical writers, who produce anonymous document sets created in a welter of collaboration among technical experts, writers, artists, editors, and users, the notion of a discourse community will perhaps seem attractive enough, but rather commonplace. They may wonder, "What's the big deal?"

But what the new thinking about discourse communities promises is no less than a descriptive model or map of typical interrelationships among writers, readers, texts, and contexts. Such a systematic explanation of how writing is produced and consumed in communities would be a valuable guide both to teachers and trainers of technical communicators and to writers venturing into new territories (new markets, new fields of expertise).

Ideally, we could elaborate this model into a full theory of contexts, an *ecology of writing* [16]. We could study this system of information exchanges much as ecologists study biotic communities, in which flora and fauna live together and interact with cycles of organic and inorganic substances. Finally, we could apply our findings to a diagnosis of problems that writers face when confronting new rhetorical situations or new information.

To date, discourse community theory has promised more than it has delivered. It is still a relatively new field of inquiry. Nevertheless, it could hold the answers to some of the most significant questions in the theory and practice of technical communication. These include questions about:

* the exact nature of collaboration,
* whether certain stylistic and graphical conventions are preferred universally or only within specific discourse communities,
* the relationship of local discourse communities (the writers and readers connected with one engineering company, for example) to global discourse communities (the discipline of electrical engineering, for example),
* the effects of incorporating new constituencies into old communities (the entrance of professional technical writers into the field of engineering communication, for example, or the entrance of home users into the market for computer products).

A social theory of communication, even in a tentative form, leads us to question and to reformulate some widely-held ideas about the writing process, many of which have never accounted sufficiently for the realities of workplace communication. Contrary to these oversimplified models of communication, discourse community theory insists that discourse is more than:

* a description of a static reality,
* the result of an author's need to put what's in the mind on the paper, and
* the transmission of a signal from an encoder to a decoder.

Discourse is the means by which communities develop and advance their agendas of action, build solidarity, patrol and extend their boundaries, and perpetuate themselves in the life of a general culture.

Local and Global Discourse Communities

In this chapter, we will explore a working definition of two kinds of discourse communities. *Local discourse communities* are groups of readers and writers who habitually work together in companies, colleges, departments, neighborhoods, government agencies, or other groups defined by specific demographic features.

Global discourse communities, by contrast, are groups of writers and readers defined exclusively by a commitment to particular kinds of action and discourse, regardless of where and with whom they work.

All writers will be involved in both local and global discourse communities. Manual writers for IBM, for example, are easily identified as members of a local discourse community—say the San Jose office, with its special requirements for researching, writing, editing, and reviewing documents. Even the whole corporation could fall within the local category, for it can be defined specifically by distinctive demographic features—number of employees, number of shareholders, places of business, names of customers, annual profits, taxes paid, and so on—and it has specific requirements for producing texts, products, and actions that represent a definite corporate culture.

Membership in a global discourse community is somewhat harder to grasp, but just as important, for it could have an equal or greater impact on how these manual writers produce documents for readers outside of, and even within, the company.

Conflict Between and Among Global and Local Communities

To expand our example, the IBM writers would find themselves working within and between two global communities, which we will define later as the *engineering community* and the *human services community.* The writers' obligations to both global communities could result in some personal dilemmas about how to proceed in their work. The engineering community tends to be oriented more toward the production side of research and development, whereas the human services community is oriented toward the consumer side. The old dichotomy of product-oriented versus user-oriented manuals arises from the conflict of these two viewpoints.

These conflicting obligations may be further complicated by conflicts between the writers' identification with the global discourse communities and their loyalty to the local discourse community. The company way of doing things may well be different from the way recommended by the global community's research and literature.

Thus writers are always negotiating among several possible frames of social identification and influence in their writing.

Recognizing and Modelling Conflict

Because of the recent trend toward empirical research in the field of writing, most work on discourse communities has involved local communities. But local communities are often shaped by the demands of global communities. Moreover, a writer's allegiance to a local community may be in conflict with an allegiance to a global community. While ethnography and other forms of empirical research support nicely the mapping of local communities, we need to draw upon social theory and large-scale rhetorical inquiry to survey the scope of global communities.

Once again, therefore, we find a need for a strong theoretical model that reaches beyond the practice of a single company or organization. As a contribution to that model, therefore, we offer the following exploration of theoretical themes in the social theory of discourse and a tentative classification of discourse communities in a technological society.

We argue that it is essential for writers and readers to understand their own relation to these larger groups and purposes. Without such knowledge, writers unconsciously submit to sometimes conflicting influences, whose effects can lead to misunderstood frustrations and difficulties in writing and social relations.

REAL VERSUS UTOPIAN DISCOURSE COMMUNITIES

As in the case of the sign, there has been a great deal of scholarly dispute over the definition of discourse communities. Joseph Harris has offered an impressively tough-minded survey of the field in his article "The Idea of Community in the Study of Writing" [11]. This article provides a good point of departure.

What Discourse Communities Are

Harris commends the work of several recent composition theorists who share an interest in sociology of writing—David Bartholomae [17], Patricia Bizzell [18, 19], and Kenneth Bruffee [13]. These writers and other "social constructionists," "sociologists of knowledge," and "ethnographical researchers" give us an alternative to cognitive models of composition by "de-mystifying the concept of intention": "rather than viewing the intentions of a writer as private and ineffable, wholly individual," they help us to see "that it is only through being part of some ongoing discourse that we can, as individual writers, have . . . points to make and purposes to achieve" [11, p. 12].

Building upon Bartholomae's assertion that "the discourse with its projects and agendas . . . determines what writers can and will do" [17, p. 139], Harris argues that "We write not as isolated individuals but as members of communities whose beliefs, concerns, and practices both instigate and constrain, at least in part, the sorts of things we can say" [11, p. 12].

The trouble with the concept of discourse communities, in Harris's way of thinking, is this: "recent theories have tended to invoke the idea of community in ways at once sweeping and vague: positing discursive utopias that direct and determine the writings of their members, yet failing to state the operating rules or boundaries of these communities" [11, p. 12]. The discourse communities of the "utopian" theorists lack material substance; they are utopias in the literal sense of being "nowhere": "In the place of physical nearness we are given like-mindedness" [11, p. 15].

Instead, Harris wants us to "reserve our uses of *community* to describe the workings of . . . specific and local groups" [11, p. 20]. In a crucial terminological argument, he claims that there are "other words—*discourse, language, voice, ideology, hegemony*—to chart the perhaps less immediate (though still powerful) effects of broader social forces on our talk and writing" [11, p. 20].

Realists and Utopians

Harris thus gives us the rudiments of a *realist theory of discourse communities* as a foil to the *utopian theory* of Bartholomae, Bizzell, and the other scholars he surveys. In the terms we suggested earlier, Harris insists on the reality only of local discourse communities and tends to discount the presence of global communities.

What would an analysis look like if it applied this realist theory to the study of communities? A number of studies have recently appeared that seem to respect the limits set by Harris. In "Discourse Communities, Sacred Texts, and Institutional Norms" [10], for example, the researchers Richard Freed and Glenn Broadhead present a classic ethnographic description of the writing practices and informing values of two consulting firms, pseudonymously called "Alpha" and "Omega," which the authors designate as "corporate cultures," each manifesting its own "discourse community."

Freed and Broadhead show how writers within these groups are subject to "cultural," "institutional," "generic," and "situational" norms that govern the production of discourse. In their view, however, the norms do not constitute the community. Instead, the community appears to be a real group of people in an actual location, the place where norms determined by "larger groups to which the community belongs" converge in special and unique ways [10, p. 163]: "Alpha and Omega are such communities; so is Freshman Composition 101; so is Jane Brown's Freshman Composition 101 class" [10, p. 162]. So here we have some examples of the "specific and local groups" that Harris thinks we should count as discourse communities.

But we can object to this way of thinking by using the same logic Harris has used to object to the utopian view of discourse communities. We have plenty of good words to describe these specific and local groups—if *culture* seems a little too high-minded, we've got *company, corporation, composition class, neighborhood.*

So why distinguish any such social group by inventing new jargon? If new terminology does not have heuristic power—that is, if it does not lead us to new theoretical or practical insights—then we should follow the law of Occam's razor, cutting away all but the most useful of philosophical terms or technical jargon and preferring ordinary language instead. New terms or metaphors should help us arrive at technical or practical information about the entity described that would not be suggested by terms already available [20].

Both the realists and the utopians, then, have taken the first steps of rhetorical inquiry (developing initial analogies and metaphors) without proceeding all the way to a working model of discourse communities. They have given examples of discourse communities and have even carried out detailed studies of these examples.

They have said what a discourse community *is like.* It is like a company or a culture or an ecosystem. But they have not gone on to define exactly what it is that these specific social groups exemplify. What distinguishes the discourse community from a discourse, a community, an ethical norm, and so on?

What Discourse Communities Do

If what discourse communities *are* is not always particularly clear, what they *do* is a little clearer. Both the utopians (Bartholomae, Bizzell, Cooper) and the realists (Harris, Freed and Broadhead) would agree on at least two points:

1. Discourse communities produce and consume discourse.
2. Discourse communities prefer certain kinds of discourse.

By classifying social groups as *discourse* communities, we focus our attention on a given group's ways of making and reading texts.

If we discover within a college, department, company, or corporation two or more general schemes for discoursing, then very likely we have located a source of conflict. This conflict may well represent the intrusion of other communities of discourse that manage to exert an influence through the medium of a single individual or group of individuals within the organization.

One of the forms of discourse will tend to be dominant and will, in a manner of speaking, fill the space of the local discourse community. Community members who prefer the other form of discourse will vie for a share of the territory and, in order to nurture their favored discourse practices, will create a theoretical space, a utopia that, in principle, would represent the ideal location for their form of discourse.

Thus arises the need for our category of global discourse communities, which are distinguished by how they discourse, not by where they are when they do it. Against Harris's objection that there are other words that work better than *community* to describe these global entities, we can argue that such words—discourse, ideology, and so on—do not have the same heuristic power. Above all, they do not suggest that global discourse communities are frequently social arrangements that serve as alternatives to groups defined by the local discourse communities of academic departments, job titles, class structure, divisions of labor, or corporate affiliations.

Different discourse communities exist within the same company, department, class, ecosystem. They may even threaten the cohesion of other social groupings. In this sense, they are indeed utopian.

Conflict Between Local and Global Discourse Communities: An Example

Say, for example, that a group of scientists has a strong commitment to basic research, an investigation into some theoretical topic that is deemed interesting only by other scientists in their own super-specialized sub-culture. They are, let us say, interested in some esoteric aspect of quantum physics.

This commitment may be at odds with the norms of the scientists' home department and university. In the local community, the administrators urge them to pursue applied research—something with definite ties to nuclear power, for example—because more grant money is available for applied research. Whereas basic research, when published, brings prestige to the university, grants bring more tangible, monetary benefits.

In resisting the demands of the university administration and by pursuing the discourse of basic research, the scientists tend to join with other scientists who have similar interests in quasi-political groups to protect and to promote their way of life. They define themselves not by a location but by a set of values and a set of characteristic actions.

Though not defined by the boundaries of space or by specific affiliations, such groups certainly demonstrate the qualities of community—hence the common phrase, the "scientific community." To protect their favored discourse paths, the basic researchers may cleave to colleagues in their specific specialties and sub-specialties (whether quark physics, herpetological ecology, or urban agricultural engineering). They may, that is, identify themselves by appeals to *disciplines*. Or they may band together with other scientists interested in the social autonomy of basic researchers. These latter arrangements correspond to commitments to global discourse communities.

Abstract Societies

Social critics of the last few decades have recognized the power of such utopian social groups. For example, in his well-known venture into social theory, *The Open Society and Its Enemies*, the philosopher of science Karl Popper speaks enthusiastically about "abstract societies," which he considers to be the primary mechanism of support for open, rational, individual action.

In his vision of how abstract societies free individuals from the demands of provincial and confining "closed societies," Popper reserves a special place for written and electronic communication and the related technologies of transportation, even though his work was written before the days of extensive computer technology [21, vol. 1, p. 174]:

> . . . an open society may become by degrees, . . . an "abstract society." It may, to a considerable extent, lose the character of a concrete or real group of men, or a system of such real groups. . . . We could conceive of a society in which men practically never meet face to face—in which all business is conducted by individuals in isolation who communicate by typed letters or by telegrams, and who go about in closed motor-cars. (Artificial insemination would allow even propagation without a personal element.) Such a fictitious society might be called a "completely abstract or depersonalized society." Now the interesting point is that our modern society resembles in many of its aspects such a completely abstract society.

Popper realizes that, despite the advent of the abstract society, "the biological make-up of man has not changed much" and that, therefore, people "have social needs which they cannot satisfy in an abstract society," but the social groups people do form (families, for example) are less effective "in the life of the society at large" than are abstract societies [21, vol. 1, p. 175].

Popper writes from a classically liberal viewpoint. According to this perspective, members of abstract societies must bear the strain created by unsatisfied emotional demands as "the price to be paid for every increase in knowledge, in reasonableness, in co-operation and mutual help" [21, vol. 1, p. 176]. But the gains of substituting abstract societies for older tribal groups far outweigh the emotional and cultural losses: "Personal relationships of a new kind can arise where they can be freely entered into, instead of being determined by the accidents of birth; and with this, a new individualism arises. Similarly, spiritual bonds can play a major role where the biological or physical bonds are weakened" [21, vol. 1, p. 175].

In Popper's "new individualism," the "abstract society" strongly influences every member's behavior and thinking; it is a normative group. But the individual remains the primary unit of action.[3]

Academic Villages

Charles J. Sykes, a recent critic of university professors, is not so sanguine about the effects of these "abstract societies" or global discourse communities. Sykes chronicles what he considers to be the negative effects upon education that result from the tendency of researchers to devote themselves more fully to their specialized fields than to the local universities that ostensibly employ them.

Borrowing Marshall McLuhan's metaphor of the "global village," he writes, "The disciplines can be thought of as academic villages, complete with elders, wise men, and elaborate rituals of initiation and ostracism. They communicate through journals, conferences, books, papers, monographs, as well as through peer review committees, professional organizations, and the form reviews of grant applications. And although they have no official hierarchy, they control the information hierarchy of status, reputation, and prestige" [22, p. 12].

Like Popper, Sykes realizes that these new social forms do not merely exist comfortably side by side with the older forms of colleges and departments, but rather compete for the professional commitment of the individuals they draw into their folds: "the rise of the villages meant a radical shift in loyalties" [22, p. 23]. According to Sykes's interpretation, by concentrating on the research agenda set by the global community, professors have lost interest in the teaching missions of their colleges: "The new power of the academic villages inevitably loosened the ties that had once bound together the various disciplines and faculties in the university" [22, p. 26]. Ultimately, professors become ghost members of the local community and commit their deepest allegiance to the global community of researchers that share the values that drive their work.

Characteristics of Global Discourse Communities

Thus, while the cosmopolitan philosopher Popper sees abstract societies as a powerful means of freeing individuals from the short-sightedness and provinciality of local groups, the conservative social critic Sykes views academic villages as elitist institutions that become powerful at the expense of the abandoned undergraduate student and the home institution [23, 24].[4]

Despite these ideological differences, the two commentators both hint at several important characteristics that should appear in any portrait of global discourse communities:

1. Global discourse communities compete with local groups for the loyalties of individual members.
2. Global discourse communities owe their existence and maintenance to high technology and to international systems of communication and transportation.

3. Both local and global communities expect their members to adhere to a set of behavioral norms and action agendas—norms and agendas that have their counterparts in the styles and genres of discourse.
4. While local communities, in addition to prescribing styles of discourse, may monitor membership by physical surveillance (corporate badges, parking stickers, correct dress, and so forth), membership in global communities tends to be regulated exclusively by discourse-governed criteria (writing style, publication in certain journals, presentations at national conventions, professional correspondence, and so forth).
5. To maintain the coherence of the group, local communities foster a romantic attachment to place (homeland, alma mater, etc.), while global communities use ideologies that appeal to Enlightenment rationalism and individualism.
6. In the second half of the twentieth century, global communities have all but totally supplanted local communities in the hearts and minds of Western intellectuals and the educated elite.

Discourse Communities in the University and the "Real World"

Since the modern university has pioneered the development of global discourse communities, we offer, in the next section, an academically based classification system from which to draw insights into how global discourse communities are organized. We risk the objection that academic writing differs distinctly from nonacademic writing and that, therefore, an academically based system of analysis has little relevance to technical communication.

As it turns out, however, the concept of global discourse communities is particularly useful in helping to show the relationship between academic and nonacademic discourse. Teachers and practitioners of technical communication too often overstate the distinction between academic writing and writing in the "real world." Certainly, there is a strong difference between writing done for and by professors (articles published in journals like *The Physical Review* or *Publications of the Modern Language Association,* for example) and writing that does not involve research specialists at all (for example, computer manuals written for home users by technical writers who have interviewed technical personnel but who have no specialized training in computer technology).

But these forms of writing represent extremes. In the context of technological action, technical communication is far more likely to involve people with a great range of specialized knowledge, from none to a great deal. As we suggested in the last chapter, business, industry, government, the military, and other institutions of American life have become ever more inextricably connected with the modern university, so much so that one critic (whom Sykes cites) has coined the phrase "university-industrial complex" to replace Eisenhower's famous designation "the military-industrial complex" [25].

If the university is connected to industry through a mutual interest in technological action, then it is reasonable to expect that the discourses of the two spheres would exert a mutual influence upon each other. And indeed this has happened.

Within the university, the need to seek funds for research activities and to market research products has greatly influenced the writing process and the kinds of writing that are done by a typical professor. In industry, the need to give work an air of professionalism leads managers, technical writers, and technicians of various stripes to connect their work with a particular academic field or discipline. The move to turn a job category like "technical writer" into a professional specialty like technical communication means the creation of a global discourse community of like-minded workers with ties to university research programs.

Far from being an absolute dichotomy, then, the distinction between academic and nonacademic writing has thus become rather subtle.

A CLASSIFICATION OF DISCOURSE COMMUNITIES

Having reasoned that professional discourse communities may be distinguished primarily by their characteristic ways of communicating, we arrive at a classification of global discourse communities by asking very simply, what do some typical texts look like?

More specifically, we might begin by asking questions that are important to most document designers in technical communication, such as the following:

- How are the texts laid out; do they contain large chunks of prose, or is the verbal material interrupted by white space, graphics, or headings?
- How long and how complex are the sentences?
- Is the language "technical" or "familiar," abstract or specific?
- What is the dominant genre and mode of discourse?
- Are the documents long and leisurely or short and dense?

Based on our pragmatic assumption that a text embodies a communicative objective, defined as the action that will result from the reading of the document, we can also ask:

- Where will the text's action be accomplished?
- What kinds of action are likely to follow the reading of the text?

Our answers to these questions, organized around tentative names of discourse communities, are summarized in Table 6 which gives a pragmatic taxonomy of discourse communities.

From the start, we should anticipate several objections by inviting readers to look upon this classification as *only one among many possible classifications*. It is not necessarily comprehensive, it is certainly not capable of explaining all local effects of global communities, and it ignores distinctions sanctioned by years of practice in university politics—above all, the division of study into disciplines like English, psychology, and physics.

Again, we focus only on the observable differences in document designs and the difference in pragmatic outcomes in preferred fields of action—what the authors in each community expect their readers *to do* after having read their documents. Our argument, then, is not that our table is comprehensive or even totally objective, but that it might be *useful* to technical communicators in sorting out the social relations they face every day.

Interpreting the Table

The logic of each column in the table of classifications may be summarized as follows.

1. The *discourse community* is a global or utopian group that substitutes for and competes with local groups and that may be further subdivided into disciplinary groups—English and history are subgroups of the humanities; biology and physics, or even life sciences, physical sciences, and social sciences are subgroups of basic science.

The generic differences in the discourse characteristics of our major groups are, of course, more pronounced than the differences between the subgroups. We are not suggesting that the fine-grained differences between the sub-groups are unimportant; we are only choosing to work with prominent discourse traits that characterize groups at a high level of generality.

These discourse utopias represent aspirations and alignments that may be invisible but nevertheless exert a powerful pull on their adherents. The same kind of reasoning with which we analyze the influence and the resulting conflicts among these general groups may be applied to an analysis of the more specific subgroups and their characteristic discourses. (Such an analysis might consider, for example, the difference in styles of documentation used by scholars in literature and history, or the different use of the word "cognitive" in the fields of cognitive psychology and educational psychology, or the different frequencies of the passive voice in the writings of electrical and mechanical engineers, and so on).

2. The *preferred field of action* describes the generic locations where the outcomes of the community's actions will be realized most directly. The field of action corresponds roughly to what rhetorical theorists usually call the *context* of the community's writing.

3. The *expected outcome of discourse* in each community describes the objective or the incipient action embodied in the discourse. The expected outcome corresponds to the *aim* of the community's writing.

Table 6. Pragmatic Taxonomy of Discourse Communities

Discourse Community	Preferred Field of Action	Expected Outcome of Discourse	Characteristic Styles and Text Designs
Humanities	the university and the "bookish world"	more discourse	unbroken blocks of text, sentences varying in length and complexity, words varying from familiar to technical, abstract to specific. Preferred genre: the essay
Basic Science (natural and social)	the university, the laboratory, and the field	more discourse (theory), scientific actions (experiment)	blocks of text broken by single- and double-order headings; varied sentence structure; high-density tables, charts, graphs; language predominantly technical, varying from abstract to specific. Preferred genre: the scientific paper, research report with theory-based problems and conclusions, closed arguments
Applied Science (natural and social)	the university, the lab, the field, industry, government	more discourse (theory), scientific actions (experiments), technological actions	similar to basic science, but with more emphasis on results; framed by practical rather than theoretical problems; open arguments

Engineering	the university, the lab, the field, industry, government, the marketplace	more discourse (theory), scientific and technological actions	medium-length blocks of text broken by multiple orders of headings; graphics varying from high to low density; language varying from technical to familiar, concrete to specific; sentences tending toward simple. Preferred genres: reports, proposals, manuals; framed by practical problems and concluding with recommendations for action
Human Services (Business, Health Professions, Applied Arts, etc.)	the university, government, industry, the marketplace	various human actions	short blocks of text with multiple headings and low-density graphics; short sentences and simple, familiar language preferred. Preferred genres: correspondence, brochures, reports, proposals, manuals, sales literature; framed by practical problem with recommendations for action prominently featured.
Art	the university, the marketplace	entertainment, enrichment	layout varies; familiar, specific language; sentences vary in length and complexity; length and form varies; graphics and language chosen for high impact. Preferred genres: poetry, fiction, drama, essay.

4. The *characteristic styles and text designs* are the conventions of writing that the community has consciously or unconsciously developed to clear the pathways to action. These characteristics constitute the *genres, modes, or styles* of the community's writing.

To giver a better idea of the (horizontal) relation of the four columns, and to see how each community relates (vertically) to the other, we will show how the special styles of discourse are suited to preferred actions in particular contexts. We must again stress that the generalizations we make about each community apply only to a utopian, or ideal, set of practices. In reality, the work of local communities and subgroups and the pressure of one global community upon another will inevitably intrude upon the highest aims of each global community.

Humanities

Compared to the typical texts of other fields, texts produced by scholars in the humanistic disciplines (languages, history, philosophy) tend to be verbally dense. The humanities trace their history to the middle ages when the preservation of verbal artifacts (manuscripts) was the chief function of scholarship. Through preservation of texts, knowledge was reproduced and extended into the future.

The destiny of the humanities still depends upon the reproduction and interpretation of knowledge that lies within a specified tradition, or *canon*, of the learning preserved in written documents.[5] *Humanistic action is the use of discourse to produce more discourse.* Literary scholars, for example, interpret selected historical texts and the writings of their colleagues in order to produce new documents that extend the tradition—literary histories, critical readings, theories of discourse, bibliographies, and revised texts of literary works.

When scholars complain that an article or book produced within the community is not useful or is without influence, they rarely mean that the book has little effect on the moral life of readers. They usually mean that the book is not successful in stimulating further discussion along its own line of argument. It is a dead end in the discourse paths of the community of interpretation.

Critics of humanistic learning—notably the Marxists and other moralists— claim that the typical writings of this discourse community are divorced from "praxis" or action. This argument privileges certain forms of action (political action, scientific action, technological action, economic action) over the favored actions of the humanist community—*reading and writing.*

Along the same lines, humanist scholars are frequently suspicious of, or ironic toward, members of their own community who claim interest in actions beyond reading and writing. Action-driven critical methods (like Marxism and pragmatism) and task-oriented subdisciplines (composition, technical writing) are generally considered marginal to the global discourse community of the humanities.

Nevertheless, no English, history, or philosophy department is ever a pure microcosm of the discourse community we are calling the humanities, which is, to use the various terms we have been developing, an "academic village," an "abstract society," a "discourse utopia," a "global discourse community."

Most English departments, for example, contain members whose discourse has more affinities with social science and thus takes on the characteristics of applied scientific discourse. Some scholars in composition studies, for instance, have worried that the rhetorical tradition provides a base of knowledge that is inadequate or ineffective in answering contemporary questions about the writing process, especially as it is practiced by "developmental" students on the fringes of literacy. These scholars have thus taken up empirical research in an effort to generate new knowledge and, in so doing, they have identified with the abstract society of applied science, or even engineering.

When such a shift in method occurs, a corresponding shift in discourse is registered. The humanist with social scientific tendencies begins to produce a discourse associated with scientific action, or experimentation. The dense blocks of text and verbal complexity of humanist discourse are signs of a community devoted to reading and writing. They presuppose a verbally dynamic author and an audience whose time can be devoted to long periods of patient reading and competent interpretation. The typical humanist author expects to be read closely and interpreted carefully. However, when a researcher's faith in the sufficiency of traditional scholarship slips, when the attraction of empiricism makes itself felt, the blocks of text began to break up, and a kind of writing emerges that enables selective reading and encourages actions that reach beyond the world of textuality.

Basic Science

Basic science is the first step away from the bookish world of humanistic scholarship. Yet it retains a commitment to tradition and a kind of canonical knowledge.

Derek de Solla Price uses the concept of "immediacy" to distinguish between the view of tradition taken alternatively in "hard science" and in the "soft sciences" and humanities. Whereas humanistic discourse ranges widely over time in its references and citations of authority, scientific articles tend to review only the most recent literature. The more recent the citations, the "harder" the science.

Price's provocative citation studies conclude that the trend to designate only recent work as part of "the literature" of science is a function of the exponential growth of scientific research. He convincingly demonstrates that "any young scientist, starting now and looking back at the end of his career upon a normal life span, will find that 80 to 90 percent of all scientific work achieved by the end of the period will have taken place before his very eyes, and that only 10 to 20 percent will antedate his experience" [24, p. 1]. Immediacy of citation, as presented in Price's work, is the best index we have of the difference between "*the* literature" of science and "literature" as it is defined in the humanistic tradition.

Basic science differs from applied science on other grounds. The discourse of basic science is structured and motivated by problems related to scientific theory, while applied research is structured and motivated by problems related to general human existence.

A basic researcher and an applied researcher may both be studying mosquitoes. But the basic researcher will be trying to answer questions about genetics or evolution, while the applied researcher will aim to generate findings that eventually could be useful in manufacturing an effective pesticide. A biology professor committed to the discourse of basic research will identify more strongly with a theoretical physicist than with a fellow biologist who does applied research.

The difference between basic and applied research is revealed in an analysis of the structure of argumentation in typical documents produced by each discourse community. The conventional genre for reporting any scientific finding, basic or applied, is the scientific paper, or research report, which generally follows the IMRaD format—Introduction, Methods, Results, and Discussion. The introduction establishes the problem to be investigated, surveys the literature reporting previous findings and arguments, and develops an hypothesis based on theory and previous findings. The methods section describes the design and the rationale of the experiment and details the steps followed to produce the desired results. The results section says what happened when the experiment or controlled observations were carried out, often presenting the analyzed findings in tables or graphs accompanied by minimal verbal texts. The discussion of the results is the interpretive section; it elaborates the analysis begun in the results section and determines the significance of the findings in the context of the overall research program of the discipline.

In using this format, basic researchers take a first step away from the discourse of the humanities by including within their writings sections devoted to reporting on and extending the scientific research program—methods and results. Nevertheless, they retain much of the traditional structure of academic argument. Like humanistic scholars, basic researchers build arguments upon the foundations of rational argument (deduction, induction, dialectic) and citation of authority (the canon of accepted masters and peer-reviewed literature), but they buttress these arguments with experimentation.

Stripped of its methods and results section, a scientific paper in basic research would look much like a humanistic essay or treatise. The aim of the introduction and discussion sections is to present and clinch a theoretical argument. The scientific portion of the paper, the methods and results, of course provides the most powerful means of clinching the argument, and is ultimately more important in distinguishing basic science from the humanities than any difference in subject matter.

Science deals with the phenomena of the natural world rather than with the phenomena of human culture, which is the province of the humanities, but this difference in subject matter has evolved as people have discovered what kinds of knowledge are best suited to the application of the scientific method. Aristotle made an important distinction when he said that practical and rhetorical reasoning is best suited to human affairs, where knowledge can never be complete because it is based on the incomplete data arising from opinion and because this data is used to answer questions about an uncertain future (such as "If we go to war, will we win?").

But it was not until the scientific revolutions of the seventeenth century that thinkers were able to make a second distinction that divided philosophical analytics—which Aristotle believed could answer questions about natural science—from empirical science. Since then, we have been able to distinguish between humanistic questions, problems of human affairs too complex to submit to the reduction required by the scientific method, and scientific questions, problems that can be solved through the controlled actions of experimentation.

The questions of basic research—questions of classification, definition, and cause and effect—were those once answered, albeit wrongly, by philosophers like Aristotle, questions of truth about the natural world, such as "Why does the earth circle the sun?" and "Why must the mosquito live one cycle of its life in water?" The methods and results sections of theoretically driven scientific discourse are thus representations of the powerful influence of the scientific method upon the history of natural philosophy.

Applied Science

The discourse of applied science takes yet another step away from the traditional discourse of the university. It applies the scientific method of investigation to questions involving human affairs, more particularly problems that relate the human realm to the natural world, questions like "What are the effects of water-soluble insecticides upon mosquito larvae?" Ultimately, the answers provided by applied science will affect nonacademic human actions and will thus attempt to close the gap Aristotle described between theoretical and practical reasoning.

However, applied science proceeds cautiously in this direction. Unlike basic science, whose discourse is characterized by confidently clinched arguments that attack or defend the claims of previous researchers, the discourse of applied science generally offers *open arguments*. Papers in applied science present findings that are fully analyzed, but they stop short of claiming to have solved the human problem related to the research.

An applied paper may show that water-soluble insecticides have some kind of limited effect upon the development of mosquito larvae, but it will leave open any questions about whether or not such insecticides should be used in a definite situation and will also avoid theoretical conclusions (such as those relating to genetics or evolutionary patterns of mosquitoes). The conclusions of applied research are objective in the sense that they are carefully limited in this way.

Whereas in basic research the methods and results sections serve the arguments set forth in the introduction and discussion, in applied research the methods and results all but stand alone, with the introduction and conclusion framing the findings by establishing the importance of the practical questions that are being asked. While the discourse of basic science produces as its favored outcome more discourse and more experimentation, the discourse of applied science foretells outcomes in the realm of technological action. It does not, however, make specific recommendations; if it did, it would have stepped over the invisible boundary between applied science and engineering.

Much of the status of applied science is caught up with its claims of objectivity, claims which depend upon its caution in drawing conclusions about the proper course of human action. Unlike engineering, applied science thus retains a residual Aristotelian suspicion about adequacy of scientific findings in determining human affairs. Applied science essentially creates a data base for others to use. The others may include basic scientists, who interpret the findings scientifically, and engineers and human services specialists, who interpret the findings practically.

Although it obviously assumes that science is an important determinant of human action beyond the laboratory and the academy, applied science leaves the work of making specific applications to engineers and policy makers who will make use of the findings it produces. In sticking close to the data and refraining from conclusions, applied science maintains at least a distant kinship with basic science, which refuses altogether to take up questions of general human action, creating an ethically and politically insulated space for research [26].[6]

What about Social Science?

Before turning to the discourse of engineering, we must face an important question: Why does social science not appear as a separate category in the table? Have we ignored the distinction between *natural* science and *social* science?

This distinction, we would argue, is based on an unusual combination of subject matter and research style rather than a set of abstract social associations of the kind we have been delineating. In fact, the social sciences are attempts to match humanist objects with scientific interpretants, to answer questions about the relations of individuals and societies by using the methods of applied science.

All of the social sciences—anthropology, sociology, psychology, political science, and economics—may appeal to a range of discourse communities and may draw upon or produce a range of discourse types:

• When theory and hypothesis-generating questions are explored in essays and informal articles, the discourse of the social sciences becomes humanistic.

The work of Karl Popper, which we quoted earlier in this chapter, is an example of social theory that deals with the subject matter of sociology but follows the historical methodology of the humanities [21]. Such writing will be read and interpreted in varying ways by sociologists, historians, philosophers, and literary scholars. Others who write in this manner include some of the most famous figures in the history of the social sciences—Karl Marx, Max Weber, Emile Durkheim, Sigmund Freud, John Maynard Keynes, to name but a few.

- When carefully operationalized experiments produce measurable results capable of being statistically analyzed, the discourse of the social sciences is identical to that of applied science. Behavioral scientists, for example, run controlled experiments that frequently involve surgical procedures done on laboratory animals and other such interventions. The research articles of these "physiological psychologists" resemble in almost every detail the writing of experimental biologists. Likewise, the work of physical anthropology shares much with sciences like geology.

- When results of the social sciences are interpreted with an eye to influencing human action, the resulting discourse relates most directly to that of engineering. Economists, for example, have often become privileged advisors in government and industry because of their success in developing statistical data into models that predict the behavior of domestic and international markets.

A pragmatic theory of discourse communities, which insists that a community be distinguished by the way its discourse reveals an interest in a special outcome, thus hints at an implicit weakness in the organization of the disciplines that distinguishes social science as a discrete field of practice [27].[7]

Engineering

Though the discipline of engineering is science-based, the discourse of engineering is far more varied than that of basic and applied science. Since, as we suggested in Chapter 7, engineering is the point of origin for the field of technical writing, it should not be surprising that the predominant genres of engineering discourse are the report, the manual, and the proposal. Moreover, since engineering itself arose from the extension of applied science into the actual field of technological work, it is helpful to think of the discourse of engineering as ranging between the poles of pure applied science, on the one hand, and human services on the other hand.

When engineers themselves write, especially research engineers in an academic setting, they tend to be most comfortable producing reports that, if not identical to the papers of applied science, add a cautious set of recommendations to the conventional IMRaD format. In many companies, especially those that emphasize research in burgeoning fields like high technology electronics or polymer development, engineers are encouraged to retain the identity of the applied scientist as represented in the discourse of caution and objectivity.[8] This ethos can be extremely useful when patent applications are made.

In other companies, especially small companies, or in departments of the large companies that do not engage in research but that are more interested in marketing, engineers produce writings whose favored outcomes are actions reaching beyond the laboratory and patent office. These are the classic texts of technical writing, of which the manual is the basic prototype, as we suggested in Part II of this book. These texts, which cluster around the human services pole of engineering discourse, are the ones most likely to be produced by technical writers, often in collaboration with engineers and marketing specialists.

Human Services

Like business, medicine, law, and many agencies of the government, engineering becomes a service profession when its main aim is to satisfy the needs of the general public in the open environment of the marketplace. In this arena has flourished the type of discourse that, in Chapter 3, we defined as the iconic-mosaic text. Though it varies significantly under the influences of local discourse communities, this general design is the norm of communication in such varied documents as users manuals, advertisements, sales brochures, direct mail, special interest magazines, how-to books of various kinds, and service or self-help pamphlets with titles like "Saving Electricity in the Home," "Living with High Blood Pressure," and "Making the Most of Your Humane Society."

With its plain style, open format, and semiotic variety, a text of this sort is geared to a readership of busy consumers who want to spend as little time as possible obtaining the information to solve a particular problem. The text is thus designed to facilitate fast and selective reading.

Art

The only one of the discourse communities whose chosen outcomes might be designated as effects upon consciousness rather than actions is the art community of poets, novelists, story-tellers, and dramatists [28].[9]

While the effects of entertainment and enrichment may seem to assume a passive readership, however, three points are worth noting in a general theory of discourse communities:

1. Art is widely acknowledged as a means by which general cultures are not only enriched but also maintained and perpetuated. Art reinforces or criticizes the values and the dreams of a culture by creating concrete images of that culture, which readers can evaluate and accept or reject as privileged signs of their communal life. In this view, entertainment and enrichment are

merely the first steps toward the more general outcome of culture-formation (or *ideology*).

2. Art may be produced in local discourse communities composed of people who have daily contact with one another (such as Emerson, Thoreau, and the Transcendentalist group of essayists and poets that gathered in Concord, Massachusetts, in the second quarter of the last century). Or the work of an individual artist may be influenced by a more abstract "school of thought" (such as "Realist Fiction" or "The Theatre of the Absurd"). In addition to the aim of culture-formation, such groups have aims similar to those of the humanities; that is, they read and write in order to read and write some more. In reading one another's work, they are encouraged, inspired, or influenced to produce work of their own. Their action agenda thus (consciously or unconsciously) involves the perpetuation of a particular kind or quality of artistic text.

3. Art stimulates commentary, discussion, critique, and interpretation. It thus activates further discourse outside of the artistic community proper in the field of literary criticism and scholarship, or humanistic action. Thus, in our profile of the actions of discourse communities, we have come full circle, beginning and ending with discourse that has further discourse as an outcome. If this were not the case, it would be tempting to set up a continuum of discourse communities that shows the progressive history of humanities yielding to technology as a move to bring discourse out of the academy and to broaden the preferred field of action. But artistic discourse, for which the readership is potentially the broadest of all discourse communities, leads back to the study of culture, inspiring rather than terminating the cycle of academic production (see Figure 41).

The artistic community thus sets limits upon the insistent movement in a technological society toward action-based discourse by providing a discourse-generating discourse in the heart of the broadest field of human action—the general public.[10] It is important to recognize art as a distinct discourse community because, while the texts of the human services community provide a present-focused image of public needs, artistic texts provide a fuller cultural context, preserving traditional needs in a canon of preferred styles and structures and hinting at the future direction of public preferences in experimental or avant garde texts.

What about Journalism?

Despite the common tendency to think of "the media" as an entity in its own right, a pragmatic theory of discourse does not grant journalism the status of a global discourse community. No doubt, the considerable influence and power of journalism, especially electronic journalism, make it a force to be reckoned with in a cultural theory of Western discourse.

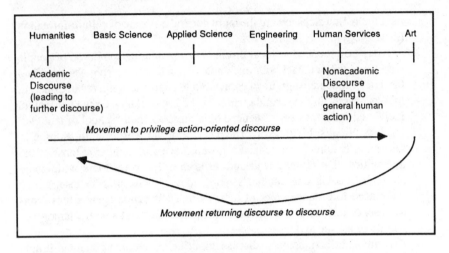

Figure 41. Continuum of discourse communities showing the trend
to distinguish academic writing from nonacademic discourse by
broadening the field of action and the trend of artistic
discourse to counteract that movement.

But we would argue that, though journalism certainly has a characteristic style of discourse, it has neither an action agenda nor a preferred field of action that distinguishes it from the work of the discourse communities. Journalists tend to contribute to the work of two discourse communities:

1. *Journalism is a service profession* in its function of providing information that a specific readership can use in making rational choices about how to act. Thus a voter reads the political news, an investor reads the business reports, and an environmentally conscious consumer reads feature articles on the effects of recycling on landfills. Journalism slides into technical communication at this point.[11]
2. *Journalism is a form of art* in its tendency to entertain its readers. Entertainment is certainly the aim of much feature journalism; and, cynics would argue, entertainment is also the main effect (if not the aim) of much news reporting.

The contrary influence of the two global discourse communities accounts for many of the conflicts within the field of journalism. Critics within and outside of the field often claim that the focus on slickness and entertainment value in news reporting is at odds with its responsibility to report reliable information, that more is spent on presentation than on substance. (This is the theme, for example, of the popular film *Broadcast News.*) The same conflict is reproduced in the literature on technical communication in the dispute, for example, over the use of brochure-manship in technical reports [29, 30].

Technical Communication as a Field of Action, Not a Discourse Community in Its Own Right

The comparison of technical communication to journalism is a telling one. Rather than being a distinct discourse community, technical communication forms as a field of practice between or among discourse communities.

As we suggested in Chapter 7, technical communication emerges as an extension of engineering into the human services field; it transforms engineering as an applied science into engineering as a service profession. This transformation, a key move in the success of technology in the general culture of the West, has a fuller context that is revealed by the history of technical communication and by the theory of discourse communities that we have been outlining.

The aims of engineering as a discipline are different from the aims of the professions that devote themselves purely to providing human services. To overcome these differences is no small task. To privilege one over the other is to court failure.

When engineering dominated technical communication in the early years, the texts that resulted were product-oriented performances that frustrated users by ignoring their needs. Now that the user-friendliness of the human service professions has been brought to bear on technical writing, the tendency could be to err in the other direction, producing texts that have the slickness and persuasiveness of the best marketing literature, but that fail to provide the information needed for effective technological actions. Technical writing involves much more than sales-manship.

An awareness of the relation of each global discourse community to the others is thus crucial in technical communication. One of the best reasons for thinking of these discourses as arising from a "discourse community" is that the community metaphor enables us to recognize that the forms of language are deeply connected to contextually motivated and aim-driven preferences for how information is to be used.

Language provides the means for expressing knowledge, and it also provides ways of marking that knowledge for particular uses in specific settings. Style limits the way information can be converted to action. Communities prefer certain styles because they prefer to use information in certain ways.

Language thus offers techniques for establishing boundaries of knowledge. Though few authors of specialized discourse would assent to the organic principle that form and content are inextricably linked, most of them act as if the principle were true. Scientists, for example, often complain that they have been misquoted by news reporters and science journalists. They may in fact have been quoted accurately, but their efforts to explain their work in a language familiar to general readers will, once rendered in print, seem strange or even altogether wrong. In an attempt to address a broad public, they have stepped over the boundaries of information set by the community within which their work is carried out.

As we suggested in Chapter 7, technical communication textbooks have tended to fossilize one kind of stylistic transformation—usually the shift from applied scientific discourse to human services discourse. No doubt, this kind of translation has occupied technical writers more than any other. But our exclusive attention to it stands in the way of a theory of technical communication that considers a broad and subtly differentiated range of discourse preferences. It tends to harden technical communication into a discourse community in its own right.

The disadvantage of this arrangement is that, though ostensibly a specialist in language, the technical communicator becomes rather a specialist in a particular kind of language, thus losing flexibility and power as a translator and a creative influence in a broad field of options for thinking, writing, and acting. Many technical writers are not simply automatons who make engineering discourse marketable. Some are bibliographers and librarians of technical information who might identify more strongly with the humanistic tradition; others are author's editors who help basic and applied scientists to communicate with one another, not with the general public; still others produce scientific journalism and thus have one foot in the community of artistic production. Very likely, a technical writer will produce all of these kinds of discourse in the space of a career.

Would it not be better, then, to envision technical communication as a set of roles and practices for facilitating exchanges among various discourse communities, a field that helps to create the possibility of an informed general public while maintaining effective links among multiple specialized communities?

The possibility of realizing this vision through effective management is the topic of Chapter 9. As we move on to that discussion, we will pursue these key points about the theory of global discourse communities:

1. Like local communities, global discourse communities develop out of specific historical conditions and thus represent traditions of past thinking and action that may be applied to human concerns in the present.
2. Global discourse communities are "abstract societies" that compete for the loyalty of their adherents within local discourse communities.
3. Like journalism, technical communication functions within and between several global discourse communities and is thus subject to divided loyalties and wide variations of practice within local communities.
4. To limit the action of technical communication to a particular discourse community (engineering or the service professions) or to a single field (the "space" between engineering and the service professions) would be to set premature and unnecessary limits on a diverse collection of communicative practices.

NOTES

[1]Peirce, drawing upon the wisdom of Socrates, pursued this line of thought in many of the same terms, creating a dialogic theory of cognition. On cognitive invention, he argued that "a person is not absolutely an individual. His thoughts are what he is 'saying to himself,' that is, saying to that other self that is just coming into life in the flow of time" [quoted in 2, page 35]. Though Peirce did not himself pursue a sociological line of thought in his general theory of signs, a number of recent social thinkers and critical theorists have made excellent use of his theory of signs in analyses that are widely skeptical about the influence of individualist psychology in communication theory. See especially the work of Eugene Rochberg-Halton [2], Robert Hodge and Gunther Kress [3], and Kaja Silverman [4]. Peirce and these commentators on the social nature of his semiotics have influenced us deeply and thus form a major part of the background of this chapter.

[2]We take up this theory at length in Chapter 10.

[3]We are indebted to Sue Sasseen for this point in our interpretation of Popper's view.

[4]An unacknowledged source for Sykes's metaphor of the academic village may be the work of Clifford Geertz, who puts forth a theory of discourse communities that is at once more ironic (distanced by the ethnographer's interpretive mood) and less influenced by the particular strains of ideology that both enliven and narrow the theories of Sykes and Popper. Geertz writes that, "when we get down to the substance of things, unbemused by covering terms like 'literature,' 'sociology,' or 'physics,' most effective academic communities are not much larger than most peasant villages and just about as ingrown. Even some entire disciplines fit this pattern: it is still true, apparently, that just about every creative mathematician . . . knows about every other one, and the interaction, indeed the Durkheimian solidarity, among them would make a Zulu proud. To some extent, the same thing seems to be true of plasma physicists, psycholinguists, Renaissance scholars, and a number of other of what have come to be called, adapting Boyle's older phrase, 'invisible colleges.' From such units, intellectual villages if you will, convergent data can be gathered, for the relations among the inhabitants are typically not merely intellectual, but political, moral, and broadly personal (these days, increasingly martial) as well. Laboratories and research institutes, scholarly societies, axial university departments, literary and artistic cliques, intellectual factions, all fit the same pattern: communities of multiply connected individuals in which something you find out about A tells you something about B as well, because, having known each other too long and too well, they are characters in one another's biographies" [23, p. 157]. In this passage, Geertz draws upon the much-cited work of Derek de Solla Price [24]. In his influential work on the history of science, Price was the first social scientist to apply Robert Boyle's term "invisible colleges" to the formation of specialized communities of knowledge in a modern setting. We have chosen to focus on the work of Popper and Sykes rather than the more measured and scholarly books of Geertz and Price because we think it is important to emphasize the *conflict* that arises between local and global discourse communities and that makes the work of technical communication both frustrating and interesting. The very lack of personal objectivity in Popper and Sykes helps them to delineate this kind of conflict.

[5]Humanists may object that the canon is currently under attack in many fields, especially literature, where there is a move afoot to admit new and more diverse texts to the accepted field of investigation. Our response is that this rebellion depends upon the previous existence of the canon. Indeed, it perpetuates it. In most cases, the argument is not to

eliminate the canon, but to broaden it. Of course, conservatives may well argue that too much breadth threatens the very existence of the canon.

[6]The shock waves that work like the Manhattan Project sends through the scientific community results from the collapsing of traditional distinctions between the discourse communities of basic science, applied science, engineering, and the military. For a fuller discussion of the distinction between basic and applied science and an analysis of some typical texts from each discourse community, see the treatment of scientific ecology in Killingsworth and Palmer [26].

[7]As an illustration of this point, consider the case of a tenured professor of physiological psychology in a small university where we used to teach. When the psychology program at the school was shut down, the man simply joined the biology department, in which he already shared a lab and a kennel of white rats with a medical researcher. Social relations between the natural sciences and the social sciences are not always so cordial, but the characteristic antagonism has as much to do with ethical and political differences as with methodological and practical differences. Natural scientists often claim that bias in social science is inevitable because social scientists are too close to the objects of their investigation—other human beings. The clarity of their experimental designs is said to suffer, and their results are considered to be clouded by emotion and ideology. See, again, Killingsworth and Palmer [26]. Also see the editorial article by J. Diamond, who writes, "These soft sciences, as they're pejoratively termed, are more difficult to study [than many of the hard sciences], for obvious reasons. A lion hunt or revolution in the Third World doesn't fit inside a test tube. You can't start it and stop it whenever you choose. You can't control all the variables; perhaps you can't control *any* variable. You can still use empirical tests to gain knowledge, but the types of tests used in the hard sciences must be modified. Such differences between the hard and soft sciences are regularly misunderstood by hard scientists, who tend to scorn soft sciences and reserve a special contempt for the social sciences" [27, p. 35].

[8]One company with which we worked, for example, insisted that all internal reports retain the convention of passive voice and omission of first-person pronouns—even the plural *we*. Not surprisingly, the company's work was on the forefront of investigation in materials engineering. The corporate identity represented a strong appeal to the global discourse community of applied science. Even high-level executives had doctoral degrees in engineering and sported the proper stylistic badge.

[9]Discourse analysts often say that art is *nonrhetorical*. For example, in his recent book on rhetorical criticism, Roderick P. Hart argues that a poet "never tells her audience exactly what she expects them to *do* as a result of reading her poem," but rather allows the audience to take "a vacation from choosing between this concrete possibility and that concrete possibility" [28, p. 6]. We would argue that the best poems are pleasurable because they create the *illusion* of this freedom from choice (whereas the best overtly rhetorical discourses are satisfying when they create the illusion of the freedom *of* choice). In fact, no discourse can be completely innocent of political, social, or cultural implications for action, though, as Hart rightly notes, some discourses can be more explicit than others about the course of action that they recommend. To put it simply, artistic discourse tends to be vague about its action agenda. We discuss the implicit rhetoricity, or instrumentality, of art in Chapter 10.

[10]In Chapter 10 we will explore the interchange of action-oriented, or instrumental, discourse and artistic discourse.

[11]No wonder that many academic programs in technical writing require courses in journalism or that terms like "science writer," which originate in the field of journalism, are in some companies applied to job descriptions of technical communicators. Colorado State University offers a well-known and widely respected Masters degree in "technical journalism." The only thing that keeps such fruitful mergers from forming at other colleges is the all too prevalent force of territorialism in departmental politics. In other words, local discourse communities interfere with the formation of links inspired by global discourse relations.

REFERENCES

1. K. B. LeFevre, *Invention as a Social Act*, Southern Illinois University Press, Carbondale, 1987.
2. E. Rochberg-Halton, *Meaning and Modernity: Social Theory in the Pragmatic Attitude*, University of Chicago Press, Chicago, 1986.
3. R. Hodge and G. Kress, *Social Semiotics*, Cornell University Press, Ithaca, New York, 1988.
4. K. Silverman, *The Subject of Semiotics*, Oxford University Press, New York, 1983.
5. G. H. Mead, *Mind, Self, and Society*, C. W. Morris (ed.), University of Chicago Press, Chicago, 1934.
6. J. Gumperz, The Speech Community, in *Language and Social Context*, P. P. Giglioli (ed.), Penguin, Baltimore, pp. 219-231, 1972.
7. J. Habermas, *The Theory of Communicative Action, Volume One: Reason and Rationalization of Society*, T. McCarthy (trans.), Beacon, Boston, 1981.
8. J. Habermas, *The Theory of Communicative Action, Volume Two: Lifeworld and System: A Critique of Functionalist Reason*, T. McCarthy (trans.), Beacon, Boston, 1987.
9. S. Fish, *Is There a Text in This Class?, The Authority of Interpretive Communities*, Harvard University Press, Cambridge, Massachusetts, 1980.
10. R. C. Freed and G. J. Broadhead, Discourse Communities, Sacred Texts, and Institutional Norms, *College Composition and Communication, 38*, pp. 154-165, 1987.
11. J. Harris, The Idea of Community in the Study of Writing, *College Composition and Communication, 40*, pp. 11-22, 1989.
12. J. E. Porter, Intertextuality and the Discourse Community, *Rhetoric Review, 5*, pp. 34-47, 1986.
13. K. A. Bruffee, Social Construction, Language, and the Authority of Knowledge: A Bibliographical Essay, *College English, 48*, pp. 773-790, 1986.
14. S. Doheny-Farina and L. Odell, Ethnographic Research on Writing: Assumptions and Methodology, in *Writing in Nonacademic Settings*, L. Odell and D. Goswami (eds.), Guilford, New York, pp. 503-535, 1985.
15. G. J. Broadhead and R. C. Freed, *The Variables of Composition: Process and Product in a Business Setting*, Southern Illinois University Press, Carbondale, 1986.
16. M. Cooper, The Ecology of Writing, *College English, 48*, pp. 364-375, 1986.
17. D. Bartholomae, Inventing the University, in *When a Writer Can't Write*, M. Rose (ed.), Guilford, New York, pp. 134-165, 1985.
18. P. Bizzell, Cognition, Convention, and Certainty, *PreText, 3*, pp. 213-243, 1982.

19. P. Bizzell, Foundationalism and Anti-Foundationalism in Composition Studies, *Pre/Text, 7,* pp. 27-56, 1986.
20. M. J. Killingsworth, How to Talk about Professional Communication: Metalanguage and Heuristic Power, *Journal of Business and Technical Communication, 3,* pp. 117-125, 1989.
21. K. R. Popper, *The Open Society and Its Enemies,* 2 vols., Princeton University Press, Princeton, New Jersey, 1966.
22. C. J. Sykes, *ProfScam: Professors and the Demise of Higher Education,* St. Martin's, New York, 1988.
23. C. Geertz, *Local Knowledge: Further Essays in Interpretive Anthropology,* Basic Books, New York, 1983.
24. D. J. de Solla Price, *Little Science, Big Science . . . and Beyond,* Columbia University Press, New York, 1986.
25. M. Kenney, *Biotechnology: The University-Industrial Complex,* Yale University Press, New Haven, 1986.
26. M. J. Killingsworth and J. S. Palmer, *Ecospeak: Rhetoric and Environmental Politics,* Southern Illinois University Press, Carbondale, 1992.
27. J. Diamond, Soft Sciences are Often Harder than Hard Sciences, *Discover,* pp. 34-39, August, 1987.
28. R. P. Hart, *Modern Rhetorical Criticism,* Scott, Foresman/Little, Brown, Glenview, Illinois, 1990.
29. R. R. Oslund, Brochuremanship—How Much is Too Much?, *STWP Review, 12*:3, pp. 14-15, 1965.
30. R. R. Oslund, Brochuremanship Versus Cost, *Data, 7,* pp. 59-60, 1962.

CHAPTER 9

Management and the Writing Process

THE WRITING PROCESS

One of the biggest contributions that contemporary research in English composition has made to the study of writing involves the analysis of the writing process, the steps by which a text comes into being. Recent review essays in the field have shown the variety and extent of this research, especially as it relates to the teaching of writing [1]. Other essays, particularly those of Richard Young [2] and Maxine Hairston [3], have indicated the importance of the new "process paradigm," which has made "process, not product" the slogan of composition teachers and which has revolutionized the teaching of writing in colleges and high schools.

The process paradigm has overturned the traditional lecture hall approach, setting up in its place writing workshops that encourage experiments in brainstorming, free-writing, journal-writing, multiple drafts, interactive and collaborative learning, revision throughout the writing process, and the use of computers to enhance individual performance and group interaction.

This treatment of writing instruction has obvious applications to the teaching of technical writing, but the implications for the workplace may be even stronger. In this chapter, we will show how discussions about the process of text production in technical communication ultimately coincide with discussions of *management*.

In 20th-century industry, management has become a kind of "people technology," an attempt to impose upon social groups structures designed to enhance productivity [4, pp. 167-320]. The organization of the writing process may be seen as a function of how technical writers and those with whom they interact are managed as they cooperate and interact in developing technical documents.

As it turns out, the two dominant views of effective management—the division of labor approach and the integrated team approach—correspond almost point for point with two competing models of the writing process, the linear model and the recursive model.

189

THE LINEAR AND THE RECURSIVE MODEL
OF THE WRITING PROCESS

Since antiquity, rhetorical theorists have been interested in the writing process. The "canons" of rhetoric, the analyzed set of operations that thinkers from Aristotle to Cicero considered essential to the creation of a good speech—invention, arrangement, style, delivery, and memory—were, in one sense, ways of institutionalizing stages or aspects of the process by which a speech is generated.

The first three of the canons have been preserved in traditional writing theory and composition pedagogy as *planning or prewriting* (invention), *drafting* (arrangement), and *revising* (style) [5-8]. This model conforms nicely to the linear version of the writing process that still prevails in many classrooms and workplaces. When writers produce a text, according to the linear model, they move in a step-wise progression from research to writing to editing and review. The research is finished when the writing begins, and editing provides the finishing touches.

Reminiscent of the old term paper assignment—in which students are urged to spend time in the library, then to turn in, in sequence, a set of note cards, an outline, a first draft, and a final draft—many text-producing operations in industry are managed according to a rigid division based on the linear model, associating the different stages of the process with different jobs: The researcher is responsible for invention; the technical writer, for drafting; the editor for revision and review.

Despite the bumper-sticker appeal of the "process, not product" slogan, the movement in composition theory that began in the late sixties was really more of a critique of the linear model than it was a shift of emphasis away from the written product. Following the major work of Janet Emig [9], the process paradigm began to assert that the writing process is not linear, but rather recursive, overlapping, and repetitive. Planning, drafting, and revising are not so easily divided into separate stages; each operation not only occurs throughout the writing process, but often supports and influences the other operations.

Most working writers can see immediately the weakness of a strictly linear view of writing. They tend to revise their work, for example, even in early stages of document production, building and developing within the process not only their sentences and paragraphs, but also their subject matter. Reviewing what they have written during a morning of writing will nearly always reveal a need for more information or a new perspective on the information. Recognizing the need, they begin the process of invention over again.

In writing a users manual, for instance, a writer may alternately take the position of the technical expert and the end user. In the latter role, adopted in early revision, the writer may discover that a key step in a procedure has been omitted. Unable to discover it without assistance—unable, that is, to fulfill the role as technical expert—the writer must seek out someone who knows more about the product before fulfilling the commitment to the end user of the manual. Thus, the writer re-views, re-searches, and revises all along.

Figure 42 is an attempt to show schematically how the linear image of the writing process differs from the recursive image suggested by recent composition theory.

Writing as Problem Solving

With the emergence of the recursive model, several distinct theories have arisen to address the question, *What motivates and controls the writing process?* The concept of *problem solving,* as developed first in design theory and cognitive psychology, has been particularly attractive to writing teachers and researchers.

In their view, problem solving takes place in all stages of composition, from brainstorming to editing, each of which incorporates such operations as preliminary observation, development of possible solutions, evaluation and testing of the working solution, and implementation of the solution [10, 11]. The theory suggests that, as writers plan, draft, and edit, they repeat again and again the stages of some (conscious or unconscious) problem-solving method. They become aware of a problem by standing back and observing their work. They develop possible

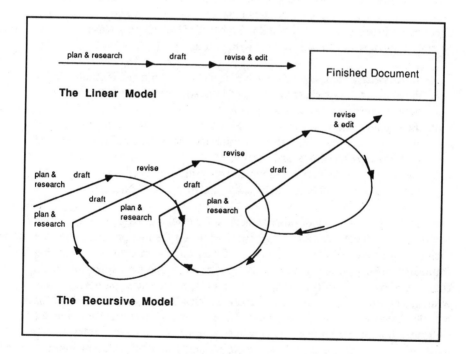

Figure 42. Two models of the writing process.

solutions and try them out, then settle on the best solution, only to face yet another set of problems. The repetition of some such procedure will occur within each recursive loop of the writing process.

Technical writing teachers have been quick to incorporate these findings into their work, noting that problem solving offers as a heuristic "a set of cumulative procedures which are flexible and recursive" [12] and which are especially useful for engineering writers who are used to applying problem-solving procedures in their technical work [13]. Not surprisingly, the authors of recent textbooks and anthologies on technical writing have enthusiastically embraced both the process paradigm and the problem-solving approach [14-16].

Competing Theories of the Writing Process

The problem-solving approach to the writing process derives largely from cognitive science. Its strong appeal in technical communication arises partly from the analogy of writing and technology in the field, which shares with cognitive psychology another analogy—between the computer and the human mind [17, 18].[1]

However, the cognitive theory of the writing process is not the only one available, as Lester Faigley has shown in an important article, "Competing Theories of Process: A Critique and a Proposal" [18]. Faigley classifies three possible approaches to the study of the writing process:

1. *the expressive view,* according to which the writing process is dominated by the individual writer's search for an "authentic voice" that faithfully represents an innermost, transcendent self,
2. *the cognitive view,* according to which the writing process may be understood as a set of problems that the writer may solve by careful analysis and step-by-step procedures, and, finally,
3. *the social view,* according to which the writing process is not only influenced but largely determined by social and historical contexts.

In considering the value of process theories for teaching, Faigley sees positive points in all of these theories, but he demonstrates a distinct preference for the social perspective, which questions the focus on the individual writer that dominated writing process theory in its first decade, when the expressionist and cognitive views were fully articulated. Moreover, in an article on nonacademic writing, Faigley questions the personalistic emphasis on audience. He explains that the social approach "moves beyond the traditional rhetorical concern for audience, forcing researchers to consider issues such as social roles, group purposes, communal organization, ideology, and finally theories of culture" [19, p. 236].

Why Apply Social Theories of the Writing Process to Technical Communication?

Social theories of the writing process are important to a theory of technical communication because they draw attention to the limitations imposed upon the process of text production by the context of production. Writers in industry often claim that textbook advice on how to arrive at the ideal document is all well and good, but constraints like time and company standards often prohibit the kinds of practice recommended.

What they are saying is that the writing process in industry is closely tied to policies and procedures as dictated by management, and that these practices severely constrain what a writer can and cannot do. Classroom teachers of writing also face a similar, though usually looser, set of constraints: A student's writing project has to be finished by the end of the term and has to meet up to a set of standards established by institutional forces external to the particular writing course.

The process must eventually produce a product, so claims for and against the process paradigm of writing instruction and practice are ultimately judged by standards of *productivity*.

Management: Writing Process Theory in an Industrial Setting

In a series of articles in the recent MLA publication, *Technical Writing: Theory and Practice,* three different commentators reveal a preference for a social theory of the writing process.

Insisting that researchers should study "the methods and processes organizations require to produce and use technical information" [20, p. 27], Roger Grice presents a model for the "information-development process," which shows how "the process used to develop technical information must mesh with the process used to develop the products that the technical information describes" [20, p. 28]. Grice's model is an eight-phase elaboration of the problem-solving paradigm that diminishes the importance of the "process, not product" slogan by showing that all products, including documents that describe other products, are considered to be the results of a cumulative, often recursive, process in an industrial setting:

1. Product-review phase
2. Objectives phase
3. Specification phase
4. Development (writing) phase
5. Verification phase
6. Production phase
7. Maintenance phase
8. Quality assessment phase [20, pp. 28-29].

Grice reinforces the view that the writing process is socially motivated and collaboratively implemented. Writing is a team activity that includes collaboration in almost every phase of production planning, drafting, and revising of trial and final documents, with participation by people whose job descriptions vary widely. In addition, he suggests that the relation of process and product, in an industrial context, appears as an issue of *management*.

The other two articles in the series take a similar view. Agreeing with Faigley and Grice that "in industry . . . the production of any text is a social process" [21, p. 36], Mary Beth Debs demonstrates the importance of organizational theory, especially the concept of division of labor, in understanding the nature of collaboration in the writing process.

The third article, by Jack Selzer, also makes use of a social perspective in an effort to understand the limits that individual writers face: "The work environment of every technical writer—whether he or she is a technical professional or a professional technical writer—obviously socializes writing activities. Every social context obviously offers freedoms and constraints" [22, p. 47]. While the work of Broadhead and Freed [23] and other students of local discourse communities [24] suggests that significant variations occur in the production process for technical documents, Selzer helpfully isolates a few general types of limits that affect the practice of almost every individual writer in the social context of industry:

- the nature of the work (research, marketing, etc.),
- a writer's position in the hierarchy of an organization,
- leadership preferences in the organization,
- availability of production technology (desktop publishing, printing, dictation machines, etc.) [22, p. 47].

What becomes clear in the work of these authors is that, when a social theory of the writing process is applied to technical communication, it dovetails with concepts of management. Like the new view of composition, the concept of management is process-oriented; and, like the concept of technology, management is concerned with *control*, with the ordering of processes to enhance productivity.

Making the Most of Competing Theories

The technological control of human beings has been strongly criticized by romantics and libertarians from Mary Shelley (the author of *Frankenstein*) to the political essayist and novelist George Orwell (the author of *1984*). These critics reject any positive value in what they feel are unnecessary constraints upon personal freedom required by a technological ordering of society.

Likewise, the application of management concepts to writing often repels theorists and practitioners with an expressivist or romantic bent (Dobrin [17], for example). They become suspicious of both the cognitive process model with its attempt to define too scrupulously the stages by which texts evolve and the social process model with its tendency to dissolve the individual communicator into networks of information exchange. In the romantic view, these approaches threaten to squelch the idiosyncratic solutions and the creativity potentially arising in the work of the unfettered individual writer.

Rather than dismissing this critique as arising from a somewhat dated and sentimental viewpoint (a theoretical move all too tempting in advanced industrial societies), we would do well to follow Faigley in his realization that all three theories of the process—the expressivist, the cognitive, and the social—have powerful features to contribute to an overall theory of management in technical communication.

In fact, the history of technical communication teaches that management has tended to become overly technical and control-oriented at the expense of personal creativity, thereby bearing out the predictions of the expressivist model. Ironically, the effort to control performance in order to enhance productivity has, over the course of this century, proved again and again to be subject to the law of diminishing returns, as the following summary of that history demonstrates.

COMPETING THEORIES OF MANAGEMENT IN TECHNICAL COMMUNICATION

Two models of collaboration have predominated in the process by which technical documents are produced. While both are concerned with productivity (for otherwise they would not be theories of management), one is a radical attempt to mechanize people's behavior, while the other represents a desire to restore a measure of freedom and flexibility to the organization of human actions. We call these the *division of labor model* and the *integrated team model.*

Division of Labor: The Good News and the Bad

Division of labor has tended to prevail since the 1950s, when "technical writer" first made its appearance as a job title. Writing, editing, and illustrating—not to mention communication management—had formerly been integrated into other activities of document production. Engineers and scientists planned, researched, wrote, and edited their own reports, for example. The addition of technical communication support personnel no doubt has improved document readability and aesthetics.

But the writing process has become steadily more fragmented and linear. In the most radical version of division of labor, certain people are in charge of planning (management), others do the research (engineers), and still others are responsible for writing (technical writers), editing (technical editors), illustrating (graphic artists), typing or word processing (clerical staff), printing and production (print shop), and packaging and distribution (marketing). The task of review and quality control returns to management, the group that has the most control of the process and the greatest opportunity to view the work as a whole.

The fragmentation of the process (from the perspective of anyone except the somewhat privileged cadre of technological managers) is further abetted by the development of specialized discourse communities—groups such as nuclear physicists, electrical engineers, systems programmers, and, now, technical communicators, who are, in turn, specializing as manual writers, newsletter and brochure managers, medical editors, and so on.

As we suggested in Chapter 8, such groups tend to distinguish themselves from others by specific discourse practices. Technical terminology, jargon, and shoptalk reinforce distinctions based on job titles or job descriptions. The insulation of these communities makes it possible for them to concentrate deeply on their given tasks with minimal interruption from outside. On this basis, the division of labor model, which has predominated in technological-capitalist society in general, contends that a concentration of specialized energy has benefits that outweigh losses due to rampant fragmentation.

By specializing in a relatively narrow task such as one stage of the text production process, workers can achieve a high degree of proficiency and a depth of expertise in the kinds of problem solving that the task generally requires. A professional manual writer, for example, becomes quickly familiar with the subtleties of second-person narration and organization of discourse units into numbered steps and chunked sections. That's the good news of division of labor.

The bad news is that the manual writer, if cut off from the process of product research and development, may be unable to participate in the process of invention in such a way as to create the best possible manual. Every writer of computer manuals has had the experience of trying to write good instructions for a sloppily constructed program. What if, for example, the logic of stepwise organization conflicts with the steps required by the software? Writers frequently are able to reveal such weaknesses as they begin to relate program capabilities to user actions. But, if research and development is rigidly divided from the process of manual drafting, if the R & D work is finished when the work of writing begins, the technical writer has no choice but to compromise the principles of good style so that the manual fits the product.

Another example: The manual writer may know that a particular block of text needs to be placed on the same page with a particular graphic. But, if the writing is finished before illustration begins, a graphic artist may lose sight of the functional relation of the verbal and visual texts and thus separate the prose and the graphic. The same thing could happen in the production stage.

Ironically, then, the specialized arts that the writer has developed because of the isolation granted by the division of labor ultimately cannot be applied if the division of labor is carried too far. The law of diminishing returns sets in.

Another problem with the division of labor is that, when people are cut off from participation in the full process of product development, they tend to lose their identification with the product as a whole and accept responsibility only for their own contribution to the product. Thus, an engineer may grieve over a bad manual for a good product. Or a manual writer will excuse a poor manual by blaming the flaws of the documentation on the yet deeper flaws of the product. And so on.

Only management—whose charge in the classic organization of division of labor is *planning, organizing, commanding, coordinating, and controlling* (review and evaluation) [25, p. 36]—will enjoy an overall image of the production process. Thus no other participant in this model is likely to view the product, the manual, the packaging, and even the advertising and marketing campaign as *an integrated whole, a package.* But this, no doubt, is the way users will view it.

The Alternative of Integrated Teams

In recognition of the shortcomings of extreme division of labor, an alternative model has begun to emerge in the technical communication industry. This approach could involve full integration of all contributors to—and even users of—a given document in every stage of the process—planning, research, writing drafts, revising, and even (given full access to computers for word processing and desktop publishing) typing and illustrating.

This alternative to the division of labor—the integrated team model—demands a new socializing of individuals from disparate discourse communities and thus represents a challenge for managers. Their main task in this approach is to oil the machinery of collaboration.

The idea of integrated teams is not a retreat to the original unity of the writing process in the individual writer. It is instead an attempt to create a good social environment that keeps the best parts of the old unified-individual concept of writing and the more recent division of labor approach.

This model admits the advantages of specialization in training, for example. Just as engineers must concentrate on a specialized field of knowledge in order to become deeply proficient, writers become better writers if they can concentrate on the finer points of language and the arts of rhetoric and text design.

But, following a period of immersion in a specialized task, the specialist should become re-integrated into a productive whole. This principle holds true whether you consider the period of immersion to be the four-years (or 6 or 8 years) of academic training or the period needed for the engineer to solve a technical design problem or the technical writer to solve a textual design problem. The student must eventually join the world of practice. The industry specialist must eventually join other specialists in an effort to "put it all together."

The integrated team model of communication management increases the opportunities of specialists to interact with one another, so that each person who will affect the overall outcome of product development can experience a good balance between periods of immersion in special problems and periods of interchange about general problems. In this way, the advantages of both the *individual-centered creative process* and the *collaborative creative process* are realized.

In the terms we developed in Chapter 8, this approach to management permits a fruitful interaction between the knowledge of *global discourse communities,* whose primary agent is the individual specialist, and the knowledge of the *local discourse community,* whose primary agent is management. By releasing some of the control of the writing process and centering it in the integrated team, company management creates a site where personal knowledge and specialized knowledge (from both global and local discourse communities) are joined in an environment that fosters yet another stage of creative work in team interaction.

The integrative model is not unique to the field of technical communication, of course, but has become in recent years a mainstay of organizational theory. The history of theory in this wider context suggests many of the finer points to be considered by managers who take up the concept of integrated teams.

HISTORY AND ORGANIZATIONAL THEORY

The roots of the integrated team model were nourished in the soil of 20th-century management theory [25]. A glance at this history will show that neither of the two models we have been discussing has yet survived a century without challenge. This fact suggests that organizational theory will continue its rapid evolution for years to come and that any predictions about which outlook toward involvement and motivation of workers will predominate are premature and risky.

Division of Labor as the First Model of Industrial Efficiency

The division of labor approach, with its origins in the classic economics of Adam Smith, was first described as an aspect of modern bureaucratic rationality in 1909 by Max Weber [26], the father of modern sociology. The Weberian analysis of bureaucratic organization recognized the following principles:

- Hierarchical distribution of power and control,
- Standardized rules and procedures,
- Specialization and division of labor according to tasks and subtasks,
- Employment based on technical competence,
- Detailed job descriptions,
- Prescriptive and rigid information flow,
- Subordination of individual needs to organizational goals.

Weber was ambivalent about the overall effects of this type of organization upon human society. In his famous historical work, *The Protestant Ethic and the Spirit of Capitalism,* he wrote cynically of the modern bureaucrat: "For the last stage of this critical development, it might well be said, 'Specialists without spirit, sensualists without heart; this nullity imagines that it has attained a level of civilization never before attained' " [27, p. 182].

Nevertheless, he was quite clear about the advantages he observed in bureaucratic rationality [26, p. 973]:

> The decisive reason for the advance of bureaucratic organization has always been its purely technical superiority over any other form of organization. The fully developed bureaucratic apparatus compares with other organizations exactly as does the machine with the non-mechanical modes of production. Precision, speed, unambiguity, knowledge of the files, continuity, discretion, unity, strict subordination, reduction of friction and of material and personal costs—these are raised to the optimum point in the strictly bureaucratic administration.

The Weberian description of efficient division of labor was supplemented by Frederick Taylor, the founder of "scientific management," also known as "Taylorism." The first "efficiency expert," Taylor contributed four new principles to the model of bureaucratic rationality [25, p. 36]:

- Workers should be selected scientifically through rigorous testing.
- Effectiveness should be objectively measured.
- Managers are planners.
- Workers are doers.

In 1916, Henri Fayol added to this theory of managerial control the five now classic principles of management: *planning, organizing, commanding, coordinating, and controlling* [25, p. 36].

Human Relations Theories as a Challenge to Bureaucratic Mechanization

As early as 1927, however, the beginnings of the integrated teams concept began to emerge as a dialectical challenge to what was by then the well established division of labor model. The founder of industrial psychology, Elton Mayo, developed the so-called "human relations approach," which offered alternatives to the system of bureaucratic management, including the following [25, pp. 38-39; 28]:

- People-oriented rather than production-oriented management,
- Use of informal work groups,
- Emphasis on cooperation rather than competition, building of community among workers and managers,
- Inclusion of workers in planning and decision-making,
- Increased concern over worker satisfaction.

Like the bureaucratic model, the human relations approach claimed to have found the key to the manager's foremost concern—productivity. The difference is that the emphasis in Taylorism and its descendants has fallen upon time and cost efficiency, whereas the emphasis in human relations management falls on *worker motivation*, which, Mayo claimed, assures the highest quality products and which is neglected and even suppressed in bureaucratic management despite any advantage that is realized in terms of time and cost.

Three recent developments within the human relations tradition support the emergence of integrated teams: the concepts of *meaningfulness of work, self-directed teams,* and *informating versus automating.*

Meaningfulness of Work

Hackman and Oldham claim that this essential ingredient in a worker's motivation—and thereby the key to increased productivity—depends upon three factors [25, p. 261; 29]:

- *skill variety*: the chance to use and develop a number of skills in a task
- *task identity*: the degree to which the worker identifies with the task, usually an effect of being involved from beginning to end
- *task significance*: the worker's recognition of the task's importance in the lives of other people within the organization and in the world of human relations in general.

There is an implicit connection between this concept and Marx and Engels' earliest critique of the alienated condition of the worker, who is forced by economic conditions to sell his laboring power but is not allowed the satisfaction of viewing the result of his labor as a whole [30]. The emphasis on meaningfulness of work also fits nicely with expressivist theories of the writing process, which stress the writer's need for personal involvement and fulfillment in the creative process.

Self-Directed Groups

Businesses have begun to experiment with project groups that work on a task from beginning to end and have a high degree of decision-making power in their project [31, 32]. This practice appears to have been influenced by the trend among Japanese managers to favor employee participation and the use of small, informally structured groups [33].

But the American system has a motive different from the tradition-sensitive managers of Japanese industry: "Given rising educational levels, demands for greater individualism, and alienation from traditional authority, it is becoming increasingly important to consider human needs" [34, p. 1188]. In one company designed to accommodate these worker needs, "members of autonomous work groups rotate jobs, select their own members, decide on assignments, monitor their own performance, provide training for each other, and are paid for the number of tasks they know how to perform" [34, p. 1189].

This approach tends toward a true democratization of work, though it is motivated by purely practical rather than idealistic interests. Since workers are better educated and expect to be treated as professionals, they must be accommodated attitudinally if companies are to realize their investments in personnel.

Rossabeth Kanter, an advocate of this approach, observes, "The organizational chart with its hierarchy of reporting relationships and accountabilities reflects one reality; the 'other structure,' not generally shown on the charts, is an overlay of flexible, ad hoc problem-solving teams. . . ." [35, p. 488]. Kanter stresses that the purpose of such groups is to increase the fertility of the setting in which ideas are generated and implemented by drawing team members from a diversity of sources and backgrounds: "It is not the 'caution of committees' that is sought—reducing risk by spreading responsibility—but the better idea that comes from a clash and an integration of perspectives" [35, p. 489].

Kanter's vision of fruitful conflict within the highly tentative unity of the group recalls the Jacksonian version of democracy popular in the early 19th century and celebrated in the work of Walt Whitman. Out of the diversity of discourse communities engaging one another with their different personal and educational backgrounds, ideas for innovation arise that would never have come about in the staid and conservative climate of the specialized in-group or in a process driven by management alone.

Automating versus Informating

In a provocative and widely read book, *In the Age of the Smart Machine* [36], Shoshana Zuboff argues that the computerization of the industrial environment creates two possible managerial options which stand in direct opposition to one another and which run parallel to the two models we have been outlining [37].[2]

According to Zuboff, managers can use computer networking either to *automate* their organization, increasing the mechanical control and impersonality of Taylorism, or to *informate* their organizations, using computer networking as a means of opening lines of communication formerly closed in the rigid bureaucratic hierarchy.

The distinction revolves around the question of *access*. If the organization chooses to limit access and control of computer resources and computer-mediated information to managers, who use the computer primarily as a tool of surveillance, gathering ever more reliable data on who is doing what when and for how long, the computerized workplace becomes more rigidly bureaucratized in its division of labor. But if, on the other hand, access to crucial information is provided to increasing numbers of people, including writers and even secretaries, then the environment is informated (or *textualized,* to use another of Zuboff's favored terms). As new vistas spread before employees whose view of the productive process had formerly been limited to the actions and decisions of their immediate sectors and offices, new opportunities for creative participation arise.

The tendency to automate, then, falls within the division of labor tradition of bureaucratic management. Only management is allowed full access to computerized information. These data are used primarily as a means of control, tracking employee performance and use of resources. The tendency to informate, on the other hand, creates a way of empowering self-directed, integrated teams. By opening computer communication to all employees, this approach contributes to the democratization of the work setting.

The Two Models Summarized

We are now in a position to summarize the two models against the background of these dialectically opposed traditions. Table 7 provides a brief comparison of the two models.

Table 7. Outline of Division of Labor and
Integrated Teams Model

Division of Labor	Integrated Teams
1. Managers hire and assign highly specialized employees to discrete roles in a closely analyzed production process.	1. Project teams composed of individuals from various specialties share in a holistically determined effort to complete a task.
2. Tasks are accomplished (i.e., documents are produced) in a stepwise, linear flow.	2. Tasks are accomplished in an informal, often recursive manner that varies according to team decisions and capabilities.
3. Process of document production is managed and reviewed "from outside" by professional managers.	3. Process is managed "democratically" by members of the team or "representatively" by a leader within the group who serves as a liaison with management.

CURRENT TRENDS IN MANAGING
TECHNICAL COMMUNICATION

The literature on the management of technical communication often takes division of labor for granted. In recent years, however, the integrated teams model has gained a number of adherents.

Division of Labor

The implicit adoption of the values of bureaucratic and "scientific" management in technical communication is suggested by the predominant concern with such issues as increased specialization, automating, and efficiency in matters of time and cost. The editors of the special issue of *Technical Communication* dealing with management echo the classic managerial values of Fayol in their claim that "information . . . is a productive resource that can be evaluated and managed for the purpose of planning, controlling, and decision-making," though they stop short of giving a mechanistic or Tayloristic definition for input (number of hours spent on a project) and output (number of pages produced) [38, p. 216]. Likewise, while reporting on managerial innovations in technical manual production, Killingsworth and Eiland [39] nevertheless rely on the Fayolian description of the manager's tasks—*planning, organizing, commanding, coordinating, and controlling*—without recognizing the possibility that management can be participatory.

In a case study of automated management, Shirley Anderson describes a "Job Tracking System" that exemplifies quite well the use of computers for the kind of authoritarian automating described by Zuboff and that thereby finds its implicit justification in the orientation and values of the division of labor model. With the system, the communication managers are able to monitor [40, p. 119]:

- Daily status of individual inputs and projects,
- Unit productivity, product cost, and vendor performance,
- Employee time entered by task.

If the classic managerial rationales do prevail in the field, how are the tasks of communicative labor divided? At least two articles have drawn on the division of the writing process into prewriting (planning and researching a topic), drafting, revising, and editing—much as it is treated in traditional composition textbooks. Krull and Hurford consider the effect on productivity enabled by computer assistance in each stage of the process [41]. Manyak encourages managerial intervention at the stages of pre-writing, revision, and editing—leaving drafting (or "scribing," to use the term of Krull and Hurford) to the individual communicator [42].

The Emergence of Integrated Teams

However, a number of authors provide case studies and anecdotal evidence suggesting that practices involving some degree of integration among discourse communities are emerging in technical communication.

Dressel, Euler, Bagby, and Dell present a system of managing proposal production that involves all members of an integrated team—proposal manager, text coordinator, writers, contract officer, text processing leader, and graphics leader—in every stage of the writing process [43]: outlining, drafting, editing, reviewing, and producing the document in its final form; moreover, this system employs informating technologies closely resembling the ideal of Zuboff.

Proietti and Thomas report on an integrated and democratically managed group for a project to develop an interactive video course involving a project manager, course developer, programmer, editor, instructional designer, consultant, funder, and subject matter experts [44].

Dilbeck and Golowich describe a company reorganization that effected a fuller integration of software developers, trainers, and technical writers, with the result that all the participating employees achieved a higher level of job satisfaction [45]. The writers attained a better sense of "the big picture," the trainers were grateful for a reduced writing load, and the developers gained a new respect for the writers and trainers through a better understanding of their functions. They also attained a higher level of job performance. Training manuals were better written, and user manuals were more technically astute.

The need to develop effective user manuals has given rise to a number of innovative efforts at collaboration and integration in the computer industry. Wendy Milner, in asserting that technical communicators can make a contribution to online documentation, seeks an expansion of the writer's role into the area of product design [46]. Barstow and Jaynes make essentially the same point in insisting on effective integration of online and hard-copy instructions for users [47]. Mark Smallwood agrees that documentation can be improved if writers are included in the initial development of software products, though he admits that communicators crossing boundaries in this way may be greeted with hostility by designers—which is predictable, given the defensive insularity demanded by the discourse communities discussed in Chapter 8— but writers can win credibility, Smallwood suggests, by achieving a degree of technical proficiency and role flexibility [48]. In the same vein, Fowler and Roeger advise close collaboration between writers and programmers [49]. Chew, Jandel, and Martinich also insist on the "enhanced role of writers in the product development cycle," especially in human factors engineering of user interfaces in computer products [50]. As early as 1983, Lee Wimberly saw this role as a means for the technical writer to gain access to a higher professional status, thus laying bare the ideological implications of integrated teams in technical communication [51].

Implementing the Integrated Teams Model

Given the theoretical advantages of integrated teams, how are technical communicators supposed to implement this new organization in practice?

While the writers of manuals, especially in the computer field, have led the way to the widespread adoption of the integrated teams model in industry, proposal writers have perhaps been most innovative in discovering the best techniques for effective integration of team members. Just as the need for good user manuals has spurred new collaborations among technical experts (programmers and designers), communicators (writers, editors, and artists), and managers in the computer industry, the need to produce a quality product under the lash of a tight deadline has prompted proposal teams to seek innovative methods of team production.

The literature on proposal writing for some years now has reported on the success of story-boarding and other graphical techniques of document planning as a way of achieving maximum input from team members at various stages of the project [52-54]. No longer considered merely a way of illustrating written points, visuals are used early in the writing process as a means of encouraging collaboration of various team members. This practice breaks the traditional linear path of document development (from planning to drafting to revising to editing to printing) and allows for overlap and recursion in the stages of the writing process. Moreover, it suggests that *an integrated process of text production can lead to a more effectively integrated, semiotically rich final text.*

Summarizing the Two Models for Managing Technical Communication

Against this background we can now refine our description of the division of labor and integrated team models to show how they work in technical communication. Table 8 gives a summary.

In the first explicit discussion of the two models of organizing technical communication, Killingsworth and Jones reported on a survey designed to determine the prevalence and effects of the two models in industry [55]. They found evidence of an increasing decentralization of the technical communication function. Where such innovations resulted in a high degree of integration, workers tended to express correspondingly high degrees of satisfaction. Surprisingly, however, little evidence emerged that pointed to an increased use of informating technologies and graphical techniques for integrating the production process.

Ideology and the Two Models

The integrated teams model appears to have great promise for several reasons, some pragmatic, some ideological. Composition teachers who have cut their teeth on the process approach to writing pedagogy will see immediately

Table 8. Division of Labor and Integrated Teams
in Technical Communication

Division of Labor	Integrated Teams
1. Only managers observe and participate in the whole process by which documents are planned and produced.	1. All members of a document team participate in all stages of production.
2. Individuals are responsible for a clearly defined task which they are trained as specialists to perform.	2. Team members are assigned roles based on their training as specialists but share and exchange roles within the team if this will improve the final product.
3. The process of writing is accomplished in a step-wise, linear fashion, with managers and technical staff planning the document, technical staff and technical writers drafting it, writers and editors revising and editing it, artists illustrating it, and word processors and printers producing final copy.	3. The stages of the writing process are overlapping and recursive, with all team members offering advice and content at every phase: writers, editors and artists, for example, may offer outlines and graphics for story-boarding during planning; and technical experts may participate in producing final copy through the use of desktop publishing software.
4. Computer technology is used to automate (monitor and manage time and costs).	4. Computer technology is used to informate (allow team members full and fast access to company files).

that the integrated teams model tends to run parallel to the image of the integrated individual author who, despite the classroom practice of a decade ago (requiring the completion of note cards, outline, first draft, and final draft all in a neatly divided, linear manner) cannot be divided into researcher, writer, and editor, but must proceed in a recursive and unpredictable direction toward the goal of completing a writing assignment.

The integrated teams model is an effort to capitalize on the advantages of the former system of integration within a single author while simultaneously capturing some of the advantages of specialization. Individual employees are still hired not as "generalists," but as specialists trained in nuclear physics, mechanical engineering, or technical writing. They therefore bring to the team depth of understanding in a particular area; the breadth necessary for full comprehension of the task of production is achieved only in the construction of the team. No one is fully in charge, and yet everyone is. The old rhetorical concern about finding a voice or a persona becomes truly a matter of corporate action, so that the image of the self as a sign in the Peircean semiotic is all the more realistic.

The politics of this approach is, as we have suggested, radically democratic, or—to use the term suggested by James Zappen [56]—pluralistic. Pluralism, which is also the political theory favored by the pragmatist philosophers William James and John Dewey, has as its primary overt purpose the maximizing of diversity within a social group. The coherence of such a group is based not upon the various individuals' adherence to preestablished norms of thought, word, or deed, but rather upon the achievement of a creative consensus each time a communal action is called for. Despite its appeal to the American mind, pluralism has been attacked by rightist and leftist ideologues and has not sufficiently addressed either critique.

Authoritarian ideologies claim that the pluralistic approach slows production and permits the "tyranny of the majority" to prevail in decision-making. The technical expert, for example, may well feel that the integrated teams model increases the status of communicators while undermining the position of the technologist, whose expertise should hold sway in design decisions. The rights of "the author"—the originator of a text, its god, as it were—are defensively asserted, for instance, against the possibility of editors working online and intruding into the document in a manner traditionally reserved for authors only. Barthes' proclamation of the "death of the author" in poststructuralist criticism [57]—the replacement of the author by the text as the chief concern of rhetorical theory—is threateningly echoed in Zuboff's call for the "textualization" of the office environment. The resistance to the use of informating technologies cited by Killingsworth and Jones may in fact stem from this concern over intrusions upon authorial (and authoritarian) territory.

On the other extreme, leftists will see the pluralistic organization of technical communication as a conservative effort to hold back a thoroughgoing socialism of the workplace. By granting increasing participation in the decision-making process to increasingly well-educated workers, management undermines the workers' class-consciousness and thereby assures their acceptance into and of the capitalist system. This form of ideological preservation of the dominant mode of production may be seen as ultimately self-threatening, however, and as likely to give way to a more thorough depowering of professional managers and technologists.

Further research is needed to determine the cultural patterns that have empowered the technical communicator at the expense of the technical expert in the new organizational structures. No doubt the rise of the consumer and service industries and the decline of hard industry have contributed to the increased demand for technical communicators and for their increased status within the information economy. It is now more necessary than ever for the scientist and engineer to market their work. Many of them claim that they spend more time writing proposals than in doing actual research. For technical experts, then, the appearance of the technical writer is a blessing and a curse—a blessing that increases the time for actual research work, a curse that threatens ever greater intrusions upon that very research.

Further Explorations

In Chapter 10, we will return to the question of consensus-formation in communicative practice. This exploration will, in turn, open up new areas of theory on the borders of our map of technical communication. Ultimately, we will find ourselves shading into questions about the overall effects of any discourse upon human action.

Turning to that broader discussion, we will need to keep these points in mind:

1. Composition theory in the last two decades has prompted a revision of theories about the writing process, leading ultimately to a substitution of a recursive model for the older linear model of planning and research, followed by drafting, revision, and editing.
2. Theories of the writing process in the field of composition overlap with concerns of management in the field of technical communication.
3. The recursive model of the writing process supports the substitution of an integrated teams model of communication management for the linearly structured division of labor model.
4. The new approach capitalizes on the advantages of both an individual-centered and a collaborative model of creativity, allowing a fruitful interaction between the knowledge of global discourse communities, whose primary agent is the individual specialist, and local discourse communities, whose primary agent is management.
5. Organizational theory suggests that an integrated teams model fosters greater worker satisfaction by allowing a fuller participation in the widest possible view of the creative process, which enhances the meaningfulness of work.
6. Ever more widely adopted in technical communication, the integrated teams model is maintained and extended through computer-prompted "informating" procedures and through semiotically diverse (especially graphical) techniques of textual invention and team production.
7. The pluralist and participatory ideology implicit in the integrated teams theory of management must survive attacks from authoritarians on the right and radical socialists on the left. It is a *liberal* and a *rationalist* ideology based in the philosophy of pragmatism.

NOTES

[1] It is precisely this manifestation of the analogy between writing and technology that David Dobrin finds disturbing in *Writing and Technique* [17], which leads him to dismiss

too quickly the conceptual connection of texts and technology, a connection that we have been exploring throughout this book. In an argument that all teachers of technical writing should heed if they are overeager to turn our field into a branch of engineering, Dobrin warns against adopting a model of the writing process that is too inflexible or mechanical, that tries to make a machine of the human mind. Lester Faigley also hints at the somewhat disturbing tendency of cognitive process models to overwork the analogy of mind and computer, turning thought into a thing [18].

²In sketching a similar set of choices, Mike Cooley, in *Architect or Bee?: The Human/Technology Relation* [37], draws upon Marxian social theory and distinguishes human labor from that of animals by granting the human designer vision of the whole, a form of imagination that allows human builders to visualize outcomes before enacting the processes that produce them.

REFERENCES

1. J. Warnock, The Writing Process, in *Research in Composition and Rhetoric: A Bibliographic Sourcebook*, M. Moran and R. Lunsford (eds.), Greenwood Press, Westport, Connecticut, pp. 3-26, 1984.

2. R. Young, Paradigms and Problems: Needed Research in Rhetorical Invention, in *Research on Composing: Points of Departure*, C. Cooper and L. Odell (eds.), NCTE, Urbana, Illinois, pp. 29-47, 1978.

3. M. Hairston, Winds of Change: Thomas Kuhn and the Revolution in the Teaching of Writing, *College Composition and Communication, 33*, pp. 76-88, 1982.

4. D. F. Noble, *American by Design: Science, Technology, and the Rise of Corporate Capitalism*, Oxford University Press, New York, 1977.

5. E. P. G. Corbett, *Classical Rhetoric for the Modern Student* (2nd Edition), Oxford University Press, New York, 1971.

6. M. J. Killingsworth, Toward a Rhetoric of Technological Action, *ITCC Proceedings, 34*, pp. RET-136-39, 1987.

7. M. J. Killingsworth, M. K. Gilbertson, and J. Chew, Amplification in Technical Manuals: Theory and Practice, *Journal of Technical Writing and Communication, 19*, pp. 13-29, 1989.

8. S. Dragga and G. Gong, *Editing: The Design of Rhetoric*, Baywood, Amityville, New York, 1989.

9. J. Emig, *The Composing Processes of Twelfth Graders*, NCTE, Urbana, Illinois, 1971.

10. L. Flower and J. R. Hayes, Problem-Solving Strategies and the Writing Process, *College English, 39*, pp. 449-461, 1977.

11. L. Flower and J. R. Hayes, A Cognitive Process Theory of Writing, *College Composition and Communication, 32*, pp. 365-387, 1981.

12. V. M. Winkler, The Role of Models in Technical and Scientific Writing, in *New Essays in Technical and Scientific Communication: Research, Theory, Practice*, P. V. Anderson, R. J. Brockmann, and C. R. Miller (eds.), Baywood, Amityville, New York, pp. 111-122, 1983.

13. A. Parker, Problem Solving Applied to Teaching Technical Writing, *The Technical Writing Teacher, 17*, pp. 95-103, 1990.

14. P. V. Anderson, *Technical Writing: A Reader-Centered Approach,* Harcourt Brace, Javonovitch, New York, 1987.
15. M. S. Samuels, *The Technical Writing Process,* Oxford University Press, New York, 1989.
16. L. Beene and P. White (eds.), *Solving Problems in Technical Writing,* Oxford University Press, New York, 1988.
17. D. Dobrin, *Writing and Technique,* NCTE, Urbana, Illinois, 1989.
18. L. Faigley, Competing Theories of Process: A Critique and a Proposal, *College English, 48,* pp. 773-790, 1986.
19. L. Faigley, Nonacademic Writing: The Social Perspective, in *Writing in Nonacademic Settings,* L. Odell and D. Goswami (eds.), Guilford, New York, pp. 231-248, 1985.
20. R. A. Grice, Document Development in Industry, in *Technical Writing: Theory and Practice,* Modern Language Association, New York, pp. 27-32, 1989.
21. M. B. Debs, Collaborative Writing in Industry, in *Technical Writing: Theory and Practice,* Modern Language Association, New York, pp. 33-42, 1989.
22. J. Selzer, Composing Processes for Technical Discourse, in *Technical Writing: Theory and Practice,* Modern Language Association, New York, pp. 43-50, 1989.
23. G. J. Broadhead and R. C. Freed, *The Variables of Composition: Process and Product in a Business Setting,* Southern Illinois University Press, Carbondale, 1986.
24. C. B. Matalene (ed.), *Worlds of Writing: Teaching and Learning in Discourse Communities at Work,* Random House, New York, 1989.
25. H. W. Cummings, L. W. Long, and M. L. Lewis, *Managing Communication in Organizations: An Introduction* (2nd Edition), Gorsuch Scarisbrick, Scottsdale, Arizona, 1987.
26. M. Weber, *Economy and Society: An Outline of Interpretive Sociology,* University of California Press, Berkeley, 1978.
27. M. Weber, *The Protestant Ethic and the Spirit of Capitalism,* T. Parsons (trans.), Charles Scribner's Sons, New York, 1958.
28. E. Mayo, *The Human Problems of an Industrial Civilization,* Macmillan, New York, 1933.
29. J. R. Hackman and G. R. Oldham, Motivation through the Design of Work: Test of a Theory, *Organizational Behavior and Human Performance, 16,* pp. 260-279, 1976.
30. K. Marx and F. Engels, *The German Ideology,* R. Pascal (ed.), International Publishers, New York, 1947.
31. J. R. Hackman, The Design of Self-Managing Work Groups, in *Managerial Control and Organizational Democracy,* B. King, S. Streufert, and F. E. Fredler (eds.), Wiley, New York, pp. 61-89, 1978.
32. E. J. Poza and M. L. Markus, Success Story: The Team Approach to Work Restructuring, *Organizational Dynamics, 25,* pp. 3-25, 1980.
33. R. Cole, Work Reform and Quality Circles in Japanese Industry, in *Critical Studies in Organization and Bureaucracy,* F. Fischer and C. Sirianni (eds.), Temple University Press, Philadelphia, pp. 421-452, 1984.
34. W. Pasmore, C. Francis, J. Haldeman, and A. Shani, Sociotechnical Systems: A North American Reflection on Empirical Studies of the Seventies, *Human Relations, 35,* pp. 1179-1204, 1982.

35. R. M. Kanter, Empowerment, in *The Leader-Manager*, J. N. Williamson (ed.), Wiley, New York, pp. 479-504, 1986.
36. S. Zuboff, *In The Age of the Smart Machine: The Future of Work and Power*, Basic Books, New York, 1987.
37. M. Cooley, *Architect or Bee?: The Human/Technology Relationship*, South End Press, Boston, 1980.
38. T. E. Pinelli, T. E. Pearsall, and R. A. Grice, Introduction to Special Issue on Productivity Management and Enhancement in Technical Communication, *Technical Communication, 34*, pp. 216-218, 1987.
39. M. J. Killingsworth and K. Eiland, Managing the Production of Technical Manuals: Recent Trends, *IEEE Transactions on Professional Communication, 29*, pp. 23-26, 1986.
40. S. A. Anderson, Managing through Automation, *ITCC Proceedings, 33*, pp. 118-120, 1986.
41. R. Krull and J. M. Hurford, Can Computers Increase Writing Productivity?, *Technical Communication, 34*, pp. 243-249, 1987.
42. T. G. Manyak, The Management of Business Writing, *Journal of Technical Writing and Communication, 16*, pp. 355-361, 1986.
43. S. Dressel, J. S. Euler, S. A. Bagby, and S. A. Dell, ASAPP: Automated Systems Approach to Proposal Production, *IEEE Transactions on Professional Communication, 26*, pp. 63-68, 1983.
44. H. L. Proietti and J. L. Thomas, Multifunctional Team Dynamics: A Success Story, *ITCC Proceedings, 35*, pp. MPD 58-60, 1988.
45. A. M. Dilbeck and L. A. Golowich, Developing the Integrated Documentation Team, *ITCC Proceedings, 35*, pp. MPD 31-33, 1988.
46. W. Milner, The Technical Writer's Role in On-Line Documentation, *ITCC Proceedings, 33*, pp. 61-64, 1986.
47. T. R. Barstow and J. T. Jaynes, Integrating Online Documentation into the Technical Publishing Process, *IEEE Transactions on Professional Communication, 29*, pp. 37-41, 1986.
48. M. S. Smallwood, Including Writers on the Software Development Team, in *ITCC Proceedings, 33*, pp. 387-390, 1986.
49. S. L. Fowler and D. Roeger, Programmer and Writer Collaboration: Making User Manuals that Work, *IEEE Transactions on Professional Communication, 29*, pp. 21-25, 1986.
50. J. Chew, J. Jandel, and A. Martinich, The New Communicator: Engineering the Human Factor, *ITCC Proceedings, 33*, pp. 422-423, 1986.
51. L. W. Wimberly, The Technical Writer as a Member of the Software Development Team: A Model for Contributing More Professionally, *ITCC Proceedings, 30*, pp. GEP 36-38, 1983.
52. D. Englebert, Storyboarding—A Better Way of Planning and Writing Proposals, *IEEE Transactions on Professional Communication, 15*, pp. 115-118, 1972.
53. J. R. Tracey, The Theory and Lessons of STOP Discourse, *IEEE Transactions on Professional Communication, 26*, pp. 68-78, 1983.
54. R. Green, The Graphic-Oriented (GO) Proposal Primer, *ITCC Proceedings, 32*, pp. VC 29-30, 1985.

55. M. J. Killingsworth and B. G. Jones, Division of Labor or Integrated Teams: A Crux in the Management of Technical Communication?, *Technical Communication, 36,* pp. 210-221, 1989.
56. J. P. Zappen, Rhetoric and Technical Communication: An Argument for Historical and Political Pluralism, *Iowa State Journal of Business and Technical Communication, 1,* pp. 29-44, 1987.
57. R. Barthes, *Image, Music, Text,* S. Heath (trans.), Hill and Wang, New York, 1977.

CHAPTER 10

The Range of
Instrumental Discourse

AN INSTITUTION AND AN AIM

Entering our tenth chapter, we might well ask, "What is Technical Communication?" We could say that it is both an institution and an aim.

As an occupation or a university course, technical communication becomes an institutional version of the processes and products of *instrumental discourse*—writing, speaking, and graphics devoted to the performance of actions, specifically technical actions. In addition to affecting job titles and course catalogs, technical communication denotes special combinations of signs and codes—*genres* like the proposal, the manual, and the report, for example, or *styles* of writing that users call "plain" or "active," or *document designs* like the "iconic mosaic." Finally, technical communication indicates special relations of authors and audiences, the technical communicator serving as a mediating link in the chain by which experts provide information that guides the action of nonexperts.

We have traced technical communication as it crystallizes into semiotic structures (Part I), genres (Part II), and communities (Part III). What we need to do now is to give a full sense of just how unstable these cultural formations are. For our theoretical model to be complete, we must show how, almost inevitably, the textual/technical crystals that we have identified dissolve back into the general field of discourse, becoming a kind of general impulse in any communication.

The Instrumental Aim in Nontechnical Discourse

This kind of impulse is what discourse theorists call an *aim*. Technical communication partakes of and represents the fullest realization of the *instrumental aim*. As manifestations of the instrumental aim, the patterns of discursive action that we call "technical" operate just as surely in genres and communities of discourse not usually associated with the electric and metal world of modern technology. From science to poetry to politics, we can discover the instrumental aim at work.

Every kind of talk and writing strives toward communicative ends. Even if not all texts are concerned with overtly "practical" needs—gathering food, making money, using computers, building dams—they are certainly devoted to bringing people together in communal relations. Even if a discourse does not always build computer networks (as in technical manuals) or political constituencies (as in campaign speeches), it usually plays some part in building *cultures*. In this sense, even seemingly non-productive texts, like poetry and fiction, may be technical. They may serve the purpose of ordering the world into productive systems.

The Instrumental Aim in Theory

The concept of instrumental discourse receives its fullest treatment in *A Pragmatic Theory of Rhetoric* by Walter Beale [1]. Beale's work represents an important achievement in discourse theory. Above all, it opens a theoretical space for documents like technical manuals and recommendation reports. In earlier taxonomies, such writings were often lumped together with news reports and scientific papers in the unwieldy category of "referential" discourse [2]. The aim of referential discourse is to portray accurately a past or present reality; it is primarily descriptive. Instrumental genres like how-to manuals and future-looking reports defy classification as referential because they are more concerned with stimulating and shaping human actions than with describing objects and reporting past events; they are *prescriptive* rather than *descriptive*.

Beale thus defines instrumental writing as "the kind of discourse whose primary aim is the governance, guidance, control, or execution of human activities. It includes such specific products as contracts, constitutions, laws, technical reports, and manuals of operation" [1, p. 94].

Instrumental texts are structured around imperative sentences (*do this*) and other performative constructions—*you (or he or she or it) will/shall/should do this*. The reader is the active agent, the "you" to whom commands, recommendations, requests, and suggestions are addressed.

The directives usually flow from a point of greater knowledge or authority to a point of lesser knowledge or authority, from a master of a technique or body of knowledge (author) to an apprentice (audience), from an expert to a non-expert. Prototypes may be found not only in the world of government and technical writing, but also in the wisdom literature and holy scriptures of ancient cultures and in many political writings. Instrumental texts are, as Beale correctly notes, situated in the gap between "contemplative" scientific writing, on the one hand, and "persuasive" rhetorical writing on the other.

But Beale does not stop at the recognition of the instrumental as a discourse aim in its own right. Though *A Pragmatic Theory of Rhetoric* is devoted above all to establishing *rhetoric* as a central and paradigmatic aim of discourse, it is the cultivation of insights about the instrumental and performative nature of all writing that makes the book *pragmatic* in the philosophical sense and that hints strongly at an extensive revision of the rhetorical as well as the scientific and the poetic aims [3].[1]

Exploring the Range of the Instrumental Aim

Encouraged by Beale's implied extension of the territory of the instrumental aim, this final chapter of our book sets out to explore the theoretical role of the instrumental in the other aims of discourse. We will argue that the rhetorical success of a document often depends upon its author's willingness to write instrumentally, to provide for readers an explicit or implicit course of action, a way to put the words and ideas to work. Though rhetoric and instrumental discourse are habitually connected in moral-religious writing (such as the *Bhagavad Gita* and the book of Proverbs) and in the characteristic literature of advanced technology (technical writing), the connection holds just as well in the discourses of theoretical science, the humanities, politics, and the institutions of democratic communication.

The first step in our argument will be to consider social and epistemological objections to an emphasis on instrumental discourse within the humanist research program—the kind of objection that stands behind the low valuation of technical communication as a topic for courses and research in English departments. The second step will be to ground the concept of the instrumental in pragmatic theories of realism and human action. The last step will be to offer some brief examples of how the instrumental aim influences action-conscious poetic and rhetorical discourse.

Instrumental Discourse and Utopia

Perhaps the most fascinating connection that emerges between technical communication and much artistic and rhetorical discourse is that these seemingly diverse forms often share a utopian drive. A poet or political visionary hopes to engage the reader in a world refashioned by the human imagination. Within a narrow range, the technical writer often does the same, with the further difference that the refashioning of the world is mediated by technology.

In a typical introduction to a manual for a computer software package, one such writer draws overtly upon utopian rhetoric: "Use this Guide while seated at your computer. Start up the program, practice with the Sample Database as you read the following chapters, and explore a whole new world of data management . . ." [4, vol. 1, p. 3]. A critical analysis of such a statement and its implied claims may well lead us to question the motives of such a text. Is this the melodramatic language of the advertiser, or do the producers of the computer technology and its manual really hope to transform the world of the user?

THE HUMANIST CRITIQUE OF INSTRUMENTAL ACTION

Despite the utopian impulse of much technical communication, perhaps even because of it, instrumentalism has been the object of a strong critique in humanist circles. We must face up to and contend with the possibility that instrumentalism has a dark side. Just as rhetoricians must ever worry about the association of their discipline with the excesses of propaganda and advertising, instrumentalists must account for the association of their characteristic discourse not only with the aesthetic dullness of the technological enterprise but also with the more sinister aspects of mechanization and social control.

Of the four forms of action that Beale associates with instrumental discourse, "governance" and "guidance" will have a familiar ring, a comfortable feel for the humanist scholar and teacher, but "control" and "execution" may raise an ugly question about the place of instrumental discourse in a humanist research program: Does Beale's introduction of the instrumental into the theory of aims represent a concession to the forces of mechanization and technologizing of knowledge that we should either discount or resist?

Weber's Concept of Instrumental Rationality

The question evokes the well known ambivalence of Weberian social theory toward the concept of *instrumental rationality,* the form of intelligence that drives the "scientific management" of human action in technological societies, the organization of people into bureaucracies.

Early in our century, in an extensive analysis of bureaucratic organization, Max Weber outlined the principles of instrumental rationality. As we indicated in Chapter 9, these principles form the basis of the division of labor approach to modern management and will thus be familiar to anyone who has worked in a government or corporate setting [5; 6, p. 34]:

- hierarchical distribution of power and control
- standardized rules and procedures
- specialization and division of labor according to tasks and subtasks
- employment based on technical competence
- detailed job descriptions
- prescriptive and rigid information flow
- subordination of individual needs to organizational goals

Ever the humanist, Weber was of two minds about the overall effects of this type of organization upon the moral life of individual human beings. On the one hand he worried about the fate of a society contained within the "iron cage" of technical bureaucracy and ruled by "specialists without spirit, sensualists without heart" [7, p. 182].

On the other hand he was well aware of the advantages he observed in instrumental rationality: "The decisive reason for the advance of bureaucratic organization has always been its purely technical superiority over any other form of organization. The fully developed bureaucratic apparatus compares with other organizations exactly as does the machine with the non-mechanical modes of production. Precision, speed, unambiguity, knowledge of the files, continuity, discretion, unity, strict subordination, reduction of friction and of material and personal costs—these are raised to the optimum point in the strictly bureaucratic administration" [5, p. 973].

The Response of Critical Theory to Instrumentalism

Thus, considering technological society at the turn of the century, Weber praised the great success of instrumentally organized social groups in increasing efficiency and productivity, in saving time, money, and energy; but he recoiled at the way instrumental organization limits the creativity and full participation of each individual.

Later in the 20th century, when Weber's interpretation of the bureaucratic means of social control was strained through the neo-Marxian perspective of the Frankfurt School of critical theory, an outright negative reading of instrumental rationality emerged. "The factors in the contemporary situation—" Max Horkheimer would write, "population growth, a technology that is becoming fully automated, the centralization of economic and therefore political power, the increased rationality of the individual as a result of working in industry—are inflicting upon life a degree of organization and manipulation that leaves the individual only enough spontaneity to launch himself onto the path prescribed for him" [8, p. 4]. "If the dream of machines doing men's work has now come true," Horkheimer cautioned, "it is also true that men are acting more and more like machines" [8, p. 26].

To extend this critique to the field of discourse would mean to compare the step-by-step procedures of a computer program to the similar structure of an ever-increasing number of instrumental and procedural documents, documents that shape ("program") the action and thought of individuals not only within government bureaus, factories, and corporate offices but also within the mass culture. Every day consumers are presented with an astounding selection of how-to books on topics ranging from home repair to computer networking, from weight loss to personal enlightenment and psychological adjustment.

Facing these manifestations in the popular literature of the fifties and sixties, the Frankfurt theorist Herbert Marcuse grew cynical over the accomplishments of American culture. He argued that "functionalization of language helps to repel non-conformist elements from the structure and movement of speech" [9, p. 86] and establishes instead an "anti-critical and anti-dialectical language," in which

"operational and behavioral rationality absorbs the transcendent, negative, opposi-
tional elements of Reason" [9, p. 97].

Marcuse thus understood instrumental discourse as a threat to both individual
creativity and social change, a reductively imperative "closed language" that
"does not demonstrate and explain" but rather "communicates decision, dictum,
command" [9, p. 101].

Communicative Action as a Corrective to Instrumental Action

So conceived, instrumental discourse supports and depends upon faith in the
rationality of an authority, a technical expert or a self-perpetuating and all-encom-
passing managerial system, whose chief competence is to establish and communicate a series of definite, specific steps that, if invariably followed, will produce a
given result in the most efficient manner possible.

Against such instrumental action, the latter-day Frankfurt School social theorist
Jürgen Habermas has developed a theory of *communicative action*. Habermas
wants to rescue a humanistically inspired version of rationality from the "iron
cage" of instrumental action. He also wants to keep social theory from falling into
the morass of pessimism revealed in the works of his forerunners in cultural
criticism (particularly Horkheimer and Adorno).

Communicative action frees rationality, as a form of goal-oriented thought and
behavior, from the absolute commitment to efficiency and hierarchical structure.
The theory of communicative action tries to redistribute the capacity for rational
action to a wider range of people with a more diverse set of goals and patterns of
thought and behavior. In terms of the management issues we raised in Chapter 9,
communicative action is the goal of integrated teams and other democratically
structured organizations, while instrumental action is the goal of so-called scientific management (Taylorism) with its rigid division of labor.

In the words of the American social theorist Mark Poster, "Instrumental
rationality [as conceived by Habermas] characterizes practices in what he calls
'the system,' that is in institutions like the bureaucratic state and the economy,
which achieve social solidarity through 'steering mechanisms'" [10, p. 23].
Instrumental rationality assigns people to definite and inflexible roles and steers
them along a highly predictable course.

Documents motivated by instrumental rationality have as their sole purpose the
control of the documents' readers. Clearly enough, such instrumental writings as
old-fashioned policies and procedures manuals and military operations manuals
create, for the purpose of maintaining the system, a narrow path of action that has
been chosen or created in advance of the document's production by hierarchically-
arranged powers. This kind of instrumental writing manifests neither the interest
in persuasion and identification that characterize *rhetorical writing*, nor the inter-
pretative and theoretical verve that characterize *scientific writing*, nor the interest

in craft, intersubjectivity, and the sharing of imaginative visions that characterize *poetic discourse*.

Narrowly instrumental documents may indeed draw upon the conventions of a democratic discourse open to information from diverse sources. And authoritarian managers may maintain the pretense of seeking "input" from various workers at all levels while intending ultimately to ignore such information when the time comes to act. However, management inspired by instrumental rationality will never treat deviant discourses with respect, but will merely note their presence, record them, and ultimately label them as "noise" in the system, which needs to be systematically ignored or expunged [11].[2]

Communicative rationality, on the other hand, "characterizes actions in what [Habermas] calls the *lifeworld*, that is, in areas of social action where socialization and cultural reproduction are at issue" [10, p. 23]. According to Habermas, "*communicative rationality* carries with it connotations based ultimately on the central experience of the unconstrained, unifying, consensus-bringing force of argumentative speech, in which different participants overcome their merely subjective views and, owing to the mutuality of rationally motivated conviction, assure themselves of both the unity of the objective world and the intersubjectivity of their lifeworld" [12, p. 10].

Effects of Instrumental and Communicative Action on the Social Construction of Reality

Both communicative and instrumental rationality are concerned with what the sociologists Peter Berger and Thomas Luckmann call the "social construction of reality" [13]. The difference lies in the means by which each form allows people to shape and intervene in this construction of their world.

Instrumental rationality tends to treat people as *objects,* as parts of the world to be manipulated and controlled. Communicative rationality treats people as *subjects,* as participants in the certification of the objective world and the construction of their lifeworld. The hierarchically-structured *organization chart* with its top-down distribution of power is the key image of instrumental rationality, while the *network or web* with its scattering of equally powerful nodes is the favored organizational model of communicative rationality.

Systems versus Lifeworlds

The *systems* of instrumental rationality are construed as a hierarchy, with experts steering other people along designated paths, but the *lifeworlds* of the subjects of communicative action are construed as the interlocking nodes of a network (or the interconnecting bubbles of a very complex—indeed undrawable—Venn diagram). Where the interests and aspirations of social subjects interconnect or overlap—encouraged by shared belief systems, historical events, good arguments, and other means of cultural identification—the possibility of *intersubjective* consensus is created.

In technological societies the liberal love of democracy, with its preference for communicative action, is challenged by the desire for efficiency, the great goal of instrumental action. The dilemma is nearly overwhelming in social situations where cultural diversity is great and increasing, as in the United States. Communicative rationality insists that, out of democratic conflict and consensus-building efforts, a reasonable course of action, an ordering of the world that is not only effective but humane, will emerge.

Instrumental Discourse with a Communicative Purpose and Structure

With the critique of instrumental reason in the background, and with a full acceptance of the desirability of a discourse that promotes communicative action, we may wonder about the place of instrumental discourse *within* communicative rationality.

Certainly, the sample documents that Beale categorizes as instrumental—contracts, constitutions, laws, technical reports, and manuals of operation—may well provide the *means* by which individuals achieve the *ends* of what Habermas has called "the ideal communication community" [14, p. 97].

Admittedly, such documents may also serve hierarchical systems of information management, may secure the locks of the bureaucratic trap, or may even advance the cause of totalitarianism (as Marcuse suggests). And yet, informed by humane ends and communicative goals, instrumental discourse may supply the realistic means by which change is accomplished.

Thus, what the social theorist Ivan Illich has suggested about technology itself is no less true of technical discourse. It can be used in two opposite ways. The first corresponds to instrumental rationality: it "leads to specialization of function, instrumentation of values, and centralization of power and turns people into the accessories of bureaucracies or machines." The second corresponds to communicative rationality: it "enlarges the range of each person's competence, control, and initiative, limited only by other individuals' claims to an equal range of power and freedom" [15, p. vii].[3]

The Example of Scientific Discourse

Consider scientific discourse, for example. Drawing upon several modes of discourse, a scientific paper is usually divided into four parts—Introduction, Methods, Results, and Discussion. The ultimate aim of the author is to sustain and advance the research/theoretical paradigm of one branch of the sciences. The community of discourse is composed of relatively equal peers, any of whom should have full access to the means of influencing the course of community action.

The precise nature of each paper's contribution to the research program as a whole is usually established in the introduction of the paper, which includes a review of the relevant literature and which suggests how the paper will falsify or substantiate previous claims. This rhetorical device establishes points of valance whereby the work of other researchers may be connected with what is reported. The results section of the paper is largely a matter of technical description, and the interpretation of those results in the discussion (or conclusion) section of the paper is mostly a rhetorical performance whose goal is to clinch the argument set forth in the introduction.

The work of clinching the argument, however, is partly accomplished by the instrumentally-oriented section that intervenes between the introduction and the results—*the methods section*. Since the "rules" of scientific discourse demand that validity rests in part upon the *reproducibility* of the results, then the methods section is a crucial element of the argument. Though it is usually narrative in structure ("substance X was exposed to Y waves for a period of Z"), the instrumental concern of "governance, guidance, control, or execution of human activities" is foremost. If, as scientific readers, we want to reproduce the experiment, we can convert the indicative sentences to imperatives with only the slightest mental effort ("expose substance X to Y waves for a period of Z").

Moreover, the narrative of procedures should conform to a "normal science" template for standard laboratory procedures, so that the results of the writer could be easily compared with results that readers have achieved in similar experiments. The discourse and the represented actions have therefore an instrumental structure. But this structure does not inhibit the communicative goals of the scientific community.

As Habermas asserts, against the objections of other critical theorists (such as Poster [10, p. 24]), "Modern science . . . [is] governed by ideals of an objectivity and impartiality secured through unrestricted discussion" [14, p. 91]. As Habermas has suggested from his earliest work, "What ultimately produces a scientific culture is not the information content of theories but the formation among theorists of a thoughtful and enlightened mode of life. . . The *empirical-analytic* sciences develop their theories in a self-understanding that automatically generates continuity with the beginnings of philosophical thought." According to Habermas, science shares with philosophy a "theoretical attitude" that allows these disciplines to free themselves from "dogmatic association with the natural interests of life and their irritating influence." This attitude derives in large part from turning logic and method into step-by-step procedures and from the habitual use of a realist discourse that ties thought to action [16, pp. 302-303].

Instrumental discourse, though ideally conditioned by normative interchanges within the communication community, emerges again and again in discourses that, like modern science, are built upon the model of communicative action. Its contribution nearly always involves a realist effort to fill the gap between mental goals (theory, consciousness) and action in the physical world (praxis, intervention). It may therefore be viewed humanistically as a kind of writing that authors use to empower readers by preparing them for and moving them toward effective action, the ends of which have been established by rhetorical, scientific, and mythic interplay in an open social context.

Summary of Theoretical Points

To summarize our main points so far:

1. The instrumental aim of discourse, the motive force of most technical communication, is devoted to the systemizing and control of human actions.
2. Though technical communication is the most obvious example of instrumental discourse, the instrumental aim may function in any form of communication.
3. Against the claims of a humanist (or a neo-Marxian) critique, instrumental discourse is not, in and of itself, oppressive. It may serve communicative ends.
4. Instrumental writing may enhance scientific and rhetorical arguments by locating a basis for the arguments in human action.

We can find a further theoretical grounding for these theses about the instrumental aim (particularly point 4) in the pragmatic conception of realism, which has been implicit in much of our discussion in earlier chapters and which we are now in a position to make explicit.

GROUNDING INSTRUMENTAL DISCOURSE IN PRAGMATIC REALISM

Philosophical realism is the basis for pragmatic thinking as first outlined by Charles Sanders Peirce and as interpreted by contemporary philosophers like Hilary Putnam and Ian Hacking. The concept of pragmatic action connects the work of these theorists with the work of Habermas and thus moves us toward a theory of instrumental discourse that includes the goals of communicative rationality [18-22].[4]

The Mind and the World Make Up the Mind and the World

The pragmatic realist begins by rejecting both the simplistic materialism that claims that *the world makes up the mind* and the simplistic idealism that claims that *the mind makes up the world*. Neither of these outlooks has much social potential: The materialist insists that meaning resides in the physical and can be mined out as well by an individual as by a community; the idealist likewise reduces the role of community by enclosing the process of meaning-making within the mind of each individual person.

The realist, as Hilary Putnam suggests, prefers an interactive model: *the mind and the world jointly make up the mind and the world* [23, p. 1]. *Practice* becomes the crucial concept in this proposition. Since the mind and the world are both subject and object, we are justified in temporarily bracketing them and moving them aside, turning our attention instead to the verb that connects them— "make up."

This is more or less the strategy of Ian Hacking. Instead of dwelling upon mind and world as a central dichotomy, he substitutes an interest in the tension between *thinking* and *doing,* both of which are versions of Putnam's "making up." In his introductory book on the philosophy of science, Hacking writes: "Science is said to have two aims: Theory and experiment. Theories try to say how the world is. Experiment and subsequent technology change the world. We represent and we intervene. We represent in order to intervene, and we intervene in light of representations" [19, p. 31].

In Hacking's view, the debate over scientific realism has been indecisive because it has focused on theory rather than practice, representation rather than intervention, and has thus produced arguments "infected with intractable metaphysics." Suspecting that "there can be no final argument for or against realism at the level of representation," he turns to *intervention.* "The final arbitrator in philosophy is not how we think but what we do," he concludes [19, p. 31], and thus develops this central maxim: "We shall count as real what we can use to intervene in the world to affect something else, or what the world can use to affect us" [19, p. 146].[5]

Pragmatic Realism in Instrumental Discourse

How does discourse work when it is developed according to this model of pragmatic realism? Realist discourse must be interventional and action-centered, or *instrumental.* It must enable action, often cooperative action between the sender and the receiver of discourse engaged in a mutual effort to effect some change in the world as they have come to know it.

Consider a famous example from ordinary language and behavior. I may or may not give an accurate representation of reality when I say, "The cat is on the mat." But, if I say to you, "Kick the cat on the mat," and you are able to do so, with the predictable outcome, then we might be able to agree on the reality of the mat, the kick, and the screeching cat. We could also say that you and I are real, and that my statement had instrumental power over the reality that we shared.

In the realist view, discourse is a tool used by real human agents to impose designs upon a real world. The world resists thought and discourse. It is a world of fact, not necessarily fact as defined by the positivist—the empirically verifiable entity prior to and separate from all thought—but fact as defined by Ludwick Fleck, "a stylized signal of resistance in thinking" [24, p. 98; 25, pp. 61-62].

Insofar as this resistance can be accounted for and then overcome, the world can be designed, controlled, and shaped by the forces of human desire, thinking, discourse, and physical conduct. This realist world is to some degree pliant, then, as are human agents themselves, who in the process of shaping the world are also reshaped. Discourse is the medium of these changes, but is also part and parcel of the world and the selves of the sender and receiver.

Language and Action

A *moving* act of discourse is generally regarded as having a strong effect on the emotions rather than the actions of the receiver. How do such discourse acts affect behavior? Can behavioral outcomes be predicted and controlled through discourse?

Charles Bazerman has noted that, even though recent studies "have demonstrated that scientific language [for example] is designed to move readers and derives from the various forces moving the authors," these studies nevertheless retain the notion of a "rift between language and the natural world": "they do not take the second step to see that mental motions influence behavior that occurs in the physical world" even though it is upon this step that "the project of empirical science is founded" [25, pp. 189-190].

A similar case is found in *speech-act theory*. Its effort to show that "a theory of language is part of a theory of action" [26, p. 17] began with J. L. Austin's famous book on the philosophy of language, *How to Do Things with Words* [27]. Austin tried to account for "performatives," utterances like "I promise" or "I bequeath" that have direct effects on the world.

But early speech-act theory, as developed in the writings of both Austin and John Searle, retreated into a series of reflections on meaning and reference, focusing on what happens on the front end of the speech act and stopping short of considering its outcomes. Ostensibly drawing on speech act theory, Michael McGee can thus write: "*the only actions that consist in discourse are performed on discourse itself.* Speech will not fell a tree, and one cannot write a house to dwell in" [28, p. 122].

Pragmatics: Goal-Oriented Speech Acts

While no one would be silly enough to suggest that an utterance could fell a tree, however, we can argue that a manual for wood-cutting represents a *virtual action* in its mediation of a conscious goal and an act that would accomplish that goal. Recent developments in the branch of speech-act theory known as pragmatics have encouraged this way of thinking, which is common sense to any writer of technical manuals and which has been treated in a number of highly suggestive articles on composition and professional writing [29-35].

Geoffrey Leech, in his general introduction to pragmatics [36], provides a model for speech acts that are *goal-oriented*, that involve *problem-solving from both the speaker's and the hearer's point of view.* Such speech acts are definitely concerned with the accurate representation of reality, but they are also directed outward in an effort to intervene in that reality.

Figure 43, based on Leech's work, shows a relation of mental, linguistic, and physical action that sums up the aim of goal-directed speech in a way that is particularly applicable to technical communication. The figure is a simple diagram of a speech act situation involving a speaker who is cold and wants the heat turned on and a hearer who is in the position to fulfill that goal. The speaker asks to have the heat turned on, and the hearer complies. From the speaker's perspective, as Leech indicates, "the problem is one of planning": What can I say to the hearer that is most likely to achieve the result I want? From the hearer's point of view, "the problem is an interpretive one": What result is the speaker trying to achieve by speaking to me in this way [36, p. 36]? The overall image of the communication situation that arises in this model is one of cooperative empowerment of agents aligned against a resisting reality.

Clearly enough, Leech's model is a simple version of the kind of communication that, in Chapter 2, we diagramed according to our Peircean scheme. Each triad of the diagram reproduced in Figure 44 contains a version of the performative speech acts analyzed by Leech.

Technical Communication as a Pragmatic Technology

Writing that neglects the instrumental needs of readers may be viewed as a product of what the American philosopher John Dewey called the "spectator

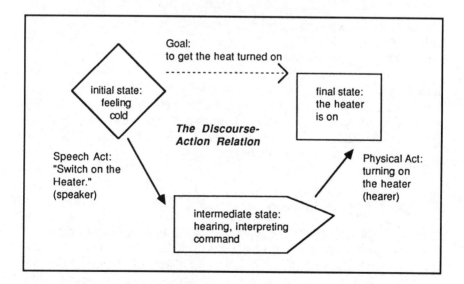

Figure 43. The goal-directed speech act (based on Leech [36, p. 37]).

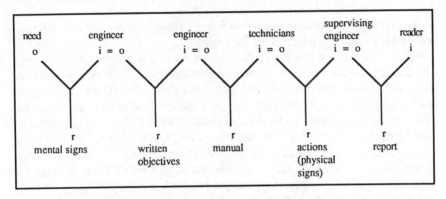

Figure 44. Semiosis in an engineering project (reproduced).

theory of knowledge" [17, p. 23]. Following Peirce's lead, Dewey developed a radical understanding of goal-oriented conduct, which he called *instrumentalism*. Dewey's instrumentalism emerged out of a dissatisfaction with spectator philosophers. He argued that, while "the realm of the practical is the region of change" [17, p. 19], the spectator theory of knowledge has tried to create a world devoid of change and its attendant contingency.

Fixing the world in a gaze, the spectator prefers discourses that universalize, that transcend change. "The depreciation of action, of doing and making," writes Dewey, "has ever been cultivated by philosophers" [17, p. 4]. The contemporary philosopher Larry Hickman has argued compellingly that Dewey's alternative philosophy of instrumentalism extends technological reasoning and action into the realm of ideas and ethics [37].

Like Dewey's instrumentalism—his "pragmatic technology"—technical communication has the smell of work and the material world. It represents the instrumentalizing of narrative. To use the words of a manual writer from IBM, the aim of the computer manual is to make the reader "the hero of a how-to story" [38]. Not only the manual, the proposal, the recommendation report—but even the sermon and the political manifesto—may draw upon instrumentalist discourse and thereby assert itself in the world of practical activity.

How is this possible?

A Pragmatic Theory of Discourse

Instead of thinking about the classification of the different kinds of writing as something established once and for all (which would be the aim of a spectator theory of discourse), we have tried in this book to take a more dynamic view, an understanding of discourse based on pragmatic aims. A similar approach, one directed toward discourse in general and not just toward technical communication (our more limited field of inquiry), is recommended in Walter Beale's *Pragmatic Theory of Rhetoric.*

Beale teaches us to think of genres, aims, and modes of writing as constantly invading one another's territory, staking new claims, and threatening to take over entirely. Thus, for example: "To the extent that [scientific discourse] attempts to influence human action and opinion, or relies for verification upon communal values, it moves in the direction of rhetoric" [1, p. 96]. Likewise: "The classical and Enlightenment conceptions of poetic art as a medium of 'instruction and delight' provide one set of terms for observing the continuum of poetic and rhetoric" [1, p. 96].

But the closest relationship of all is found between rhetorical and instrumental discourse: "to the extent that [rhetoric] takes on the capability of directing human activities . . . , it becomes instrumental" [1, p. 96], and "to the extent that [instrumental] discourses amount to *recommendations* of . . . activities . . . , or recommendations of procedures in competition with other procedures, they move in the direction of rhetoric" [1, p. 95-96].

The theory rests on a semiotic concept that Beale calls the "motivational axes," a diagram of which is given in Figure 45.

Beale says that none of the four aims identified in the model—scientific, instrumental, rhetorical, or poetic—can ever be anything other than *performative.* He thus offers a radical critique of the concepts of referential discourse, for example: "the view of meaning-as-reference tends to obscure the fact that discourse is meaningful and successful not merely to the extent that it *refers to realities* correctly or effectively but also to the extent that it *performs actions* in an appropriate and satisfactory way" [1, p. 91]. As Rorty suggests in a similar application of Dewey's pragmatism, "the notion of 'accurate representation' is simply an automatic and empty compliment which we pay to those beliefs which are successful in helping us *do what we want to do*" [39, p. 10; italics added].

Like other pragmatists, Beale has a special understanding of the idea of *aim*: "the aims constitute identifiable *norms of activity,* inevitably involving not merely psychological and linguistic norms but also *the values of community*"; moreover, the aims must be understood as "very much historical products, *developing and changing in time*" [1, p. 88; italics added].

To some extent, then, in the spirit of Dewey and despite a central allegiance to rhetoric, Beale posits an instrumentalism at the heart of all discourse. Instrumental writing is merely the prototype for the instrumental aim in all communication.[6]

The instrumentalizing of discourse suggests that the lowly technical manual—whose overt goal is either to control or to empower readers by altering their behavior—is a kind of discourse to which others aspire to some degree. The success of rhetorical, poetic, and scientific discourse may well be determined by the degree to which they make their influence felt instrumentally in the world.

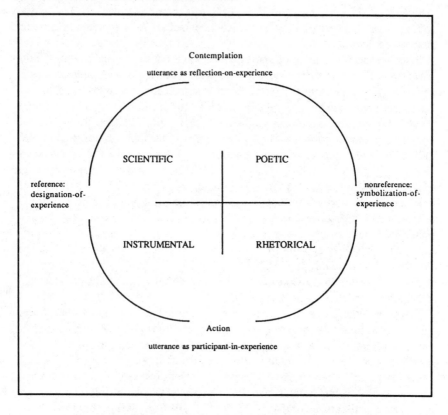

Figure 45. The motivational axes of discourse (based on Beale [1, p. 11]).

THE INSTRUMENTAL AIM IN THE LITERATURE
OF PERSONAL EMPOWERMENT

The entry of humanist scholars into the field of technical writing has coincided with a refashioning of the technical manual that accommodates aims usually associated with rhetorical and poetic texts—above all, a concern with audience-conscious and intersubjective communication.

Several authors, including Merrill Whitburn [40], David Bradford [41], and Brenda Rubens [42], have advocated a new personalism in the production of the technical manual, for example—a reliance on a strong persona or on the cultiva-tion of an ethos of helpful encouragement—as a means of meeting the needs of contemporary users of home technologies like the microcomputer. In Chapter 6 we drew upon late structuralist theory to modify this approach and to shift the interest of technical communicators to the process by which readers and writers are textualized in the production and use of technical manuals.

The rhetorical approach of writers like Whitburn, Bradford, and Rubens and our own semiotic approach have in common an interest in the *empowerment of readers*, extending the range of a technology's usefulness, making tools "convivial," which is the great goal of the interaction of democracy with "human-scale technology" as envisioned by writers like John Dewey [17], E. F. Schumacher [43], Lewis Mumford [44, 45], and Ivan Illich [15].

Having brought their discipline to bear on technical communication, however, rhetorical scholars would do well to trace the presence of the technical impulse, or instrumental aim, in documents whose primary goals are artistic or rhetorical. A full dialogue between the technical and the humanistic, between means-oriented and ends-oriented discourses, could prepare the way for an expanded and refreshed version of communicative action in the humanistic disciplines in general and in English departments in particular.

We have already shown how the instrumental aim, the heart of technical communication, is built into the scientific discourse as well. We can now consider briefly how the aim may also inform the poetic and rhetorical discourses of democratic culture in America.

Instrumentalizing Poetry:
Walt Whitman's Manual of Selfhood

Our first example comes from the writing of a mid-19th-century journeyman carpenter and printer from Brooklyn—Walt Whitman. The work in question is a poem, but, if we omit the exclamation points, these lines could be mistaken as instructions from a carpentry manual [46, p. 50]:

> Unscrew the locks from the doors!
> Unscrew the doors themselves from their jambs!

Of course, most readers do not begin dismantling their doorways upon reading these lines. But, if the aim of the poem is to move the reader toward some kind of action (albeit an action not altogether clear in the poem itself, a "protopolitical" action), then our bringing together of "Song of Myself" and the technical manual is not as gratuitous as it may seem.

Whitman's "Song" is not about *being* a self but about *becoming* a self, and it is as much about the reader as about the persona. Scholars in technical writing have drawn upon poetic art in suggesting that the presence of a strong persona enhances instrumental documents by providing the reader with the illusion of a reliable personal guide to help with the task at hand [40-42]. The immediate context of the lines just quoted reveals that Whitman would agree with this approach to discourse. In fact, he coined the term *personalism* to describe his approach to writing ([46, p. 50]; ellipses in original):

> Walt Whitman, an American, one of the roughs, a kosmos,
> Disorderly fleshy and sensual . . . eating drinking and breeding,

> No sentimentalist . . . no stander above men and women or apart from
> them . . . no more modest than immodest.

> Unscrew the locks from the doors!
> Unscrew the doors themselves from their jambs!

"Song of Myself" is a manual of selfhood devoted to the empowerment of the
reader. Just before the lines just quoted appears this passage [46, pp. 49-50]:

> I am less the reminder of property or qualities, and more the reminder of life,
> And go square for my own sake and for others' sake,
> And make short account of neuters and geldings, and favor men and women
> fully equipped,
> And beat the gong of revolt, and stop with fugitives and them that plot and
> conspire.

Written in the early 1850s, in the period of exhilaration following the Revolutions
of 1848, "Song of Myself" is a celebration of revolutionary hope. But it is more
than a celebration, a backward looking end-in-itself; it is an active effort to extend
the French and American Revolutions, to bang the gong of revolt until freedom is
a realization rather than an ideal.

Much like the most popular American novel of the day, Harriet Beecher
Stowe's *Uncle Tom's Cabin*—which, in the estimation of Abraham Lincoln, may
be credited with starting the war against slavery—"Song of Myself" aims
toward an active intervention in the lives of real people. The poem is designed to
move people to action. Not satisfied with making art, Whitman wanted to make
history [47].

Most humanist readings of "Song of Myself" are based on an idealist
philosophy, which focuses on *agent* rather than *act* [48]. The primary effects of
the poem are located in the life or mind of the author, as the genetic schools of
criticism would have it; or the effects are located in the poem itself or in the world
of discourse, as the textualists would have it, substituting text for poet or persona
but remaining faithful to the idealist privileging of agent.

In these readings the author or the mind or the text is located as the source of
power, but what of the reader? An instrumentalist or pragmatic reading would
shift our attention to the effects of the words on the reader's world of action. It
would ask, How does reading the poem affect the ability and the inclination of the
reader to intervene in history?

Insofar as the poem affects the *inclination* to act, it is rhetorically effective.
Insofar as it affects the *ability* to act or the *kind and quality* of the action, it is
instrumentally effective. Just as computer manuals open new worlds of possibilities for readers, so *Uncle Tom's Cabin* and "Song of Myself" provide models
of action for inspired readers.

Instrumentalizing Rhetoric:
King's Manual of Right Action

The shift in critical interest away from effects in the mind and work of the author and toward effects in the world and action of the reader, a shift toward pragmatic realism, is not only characteristic of ideological, new historicist, and reader response criticism, but also of the aesthetic favored by African-American writers, of whom the rhetorical analyst Reginald Martin writes: "Certainly, to critically appraise a book as to whether it 'helped' anyone live a better life would repel Formalist, New Critic, and Structuralist critics alike. Yet motivating others to begin thinking a better, humane way of living is one of the initial aims of Western humanism, and this motive is also clearly the aim of the new black aesthetic criticism" [49, p. 380; 50, pp. 19-40].

Martin rightly seeks the origins of the new black aesthetic not in any particular poetic movement, but in the work of the social reformer Martin Luther King, Jr. He focuses particularly on King's famous letter from the Birmingham jail [49, pp. 374-375]. It is unlikely that this letter, often included in rhetoric readers, has been read in our classrooms as an intensely instrumentalist document. But King's words are nothing if they are not "moving"; and instrumentalism, we would argue, secretly informs the popular notion of what constitutes a moving discourse.

Here, for example, is King's critique of "the strangely irrational notion that there is something in the very flow of time that will inevitably cure all ills" [51, pp. 859-860]:

> Actually, time itself is neutral; it can be used either constructively or destructively. More and more I feel that the people of ill will have used time more effectively than have the people of good will. We will have to repent in this generation not merely for the hateful words and actions of the bad people, but for the appalling silence of the good people. Human progress never rolls in on wheels of inevitability; it comes through the tireless efforts of men willing to be co-workers with God, and without this hard work, time itself becomes an ally of the forces of social stagnation. We must use time creatively, in the knowledge that the time is always ripe to do right. Now is the time to make real the promise of democracy and transform our pending national elegy into a creative psalm of brotherhood. Now is the time to lift our national policy from the quicksand of racial injustice to the solid rock of human dignity.

In this passage we find the artful amplification of basic instrumental imperatives, a recipe for altering consciousness and action:

- Use time as an instrument for good,
- repent of your silence,
- work hard,
- open the doors of democracy,

- recognize the dignity of all human beings, and
- put an end to racism in America.

History has shown how such words mobilized a real community of real suffering people and real sympathizers. Such words intervened in and altered the reality of the writer and his readers. Moreover, King's ends-oriented discourses were accompanied by workshops on the techniques of nonviolent protest, a methodology that the Civil Rights movement established as an effective means of change in American social history.

The very success of King and his movement may be attributed to this powerful synthesis of means and ends, which corresponds to the synthesis of form and content in the literary efforts of Whitman (free verse about the freedom of the autonomous self) and in the works of African-American poets and novelists in 20th-century America (strong words for strong themes), who carry forward the literary tradition of Stowe, Whitman, Frederick Douglass, and King, all writers interested in intervening in social and political history as well as literary history.

The Cultural Perspective on Technical Communication

If not the intensity, then certainly the concreteness of a moving discourse derives from an instrumentalist foundation with surprising frequency. The concreteness *and* the intensity partly arise from the tendency of certain genres to favor a specific orientation in time and a specific relation of author and reader.

Instrumental writing aims to get work done, to change people by changing the way they do things. Regardless of the rational or rhetorical framework within which the instrumental aim operates, it embodies an overwhelming need to create an action agenda and to prepare readers to carry it out. Just as storytelling provides traditional communities with a vehicle for purveying time-honored truths, instrumental discourse performs a crucial function in communities that rely upon method as a means to establishing new truths.

In technical writing, the report shapes information from the past, the proposal carries forth ideas into the future, and the manual makes special claims on the present activity of the reader. All of the genres combine verbal and visual signs in an effort to engage the reader in a world-shaping technology. Technical communication is thus an essential tool of *Homo faber,* the human doer or maker, the active technologist.

This larger, cultural perspective allows us to add a new layer to the three-part theory of technical communication we first introduced in Chapter 1. The revised model is given in Figure 46.

Traditionally, textbooks and researchers in the field have approached technical communication through genre and style. The more recent concern with ethnography and design broadens the field to encompass studies at the broader cultural level and the more basic semiotic level.

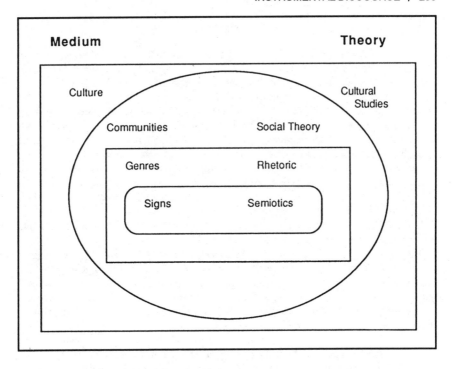

Figure 46. Signs, genres, and communities in the full cultural context.

In support of these recent developments in the field, our theoretical musings indicate the following interconnections among the four layers of the diagram in Figure 46:

1. The styles and genres of technical communication represent combinations and extensions of more basic semiotic processes—in particular the processes of representation and interpretation.
2. The styles and genres also represent crystallizations and concentrations of community preferences for discourse and action.
3. These communities in turn represent specialized functions within a larger cultural framework.
4. Finally, the cultural framework itself can only be comprehended through the semiotic processes of representation and interpretation. The largest container is thus itself contained.

In conclusion, we can say that the four layers of study and practice are interconnecting and interpenetrating. Each level affects all the others in the processes by which technical communicators represent and interpret thoughts, texts, and actions in the technological world.

NOTES

[1]A focus on the instrumental aim urges us toward a new understanding of the relationship between discourse and other human actions. This relationship is crucial not only to pragmatic semiotics but also to many socially conscious movements in the field of composition, including literacy and feminist studies, as well as two of the three epistemological perspectives that James Berlin describes in his monograph on the recent history of writing instruction—the "subjective" and the "transactional" (only the "objective" stance is not action-oriented) [3]. Thus, just as surely as technical writing teachers and practitioners of technical communication will need to take account of a discourse aim that is directly concerned with the "governance, guidance, control, or execution of human activities," so will ethnographers, social constructionists, cognitivists, historicists, expressivists, ideologists—all members of schools of research and theory who are interested in the ways and means of human action and empowerment.

[2]See Killingsworth and Palmer [11], Chapter 5, for examples of quasi-communicative instrumental actions in government discourse.

[3]Habermas allows so much himself. In his introduction to *The Theory of Communicative Action,* he begins by following the tradition established by the Frankfurt School in dividing "instrumental mastery" with its "realistic" approach from "communicative understanding" with its "phenomenological approach." But, after showing how both approaches contain the seeds of rationality, that both depend upon "the reliability of the knowledge contained" in their characteristic expressions, Habermas departs from his predecessors in arguing that "The concept of cognitive-instrumental rationality can be fit into [the] more comprehensive concept of communicative rationality developed from the phenomenological approach" [12, p. 8-14]. Since phenomenology posits the need for an intersubjectivity beyond mere subjectivity, one avenue to agreement or consensus may well be that staked out by the realism of instrumental rationality. The dialectic could be summarized thus: the conjunction of the thesis "subjectivity" with its antithesis "instrumental rationality" produces the synthesis "intersubjective or communicative rationality."

[4]Since this version of pragmatism is essentially opposed to the nominalist strain that runs from William James down to Richard Rorty, we have chosen to focus not on pragmatism per se but on the concept of realism, which was a favorite of Peirce's because of his interest in scholastic philosophy and which has been taken up by Putnam and Hacking. "The real" was also a major concern of Karl Marx, who, like Peirce, felt that meaning was a function of practical outcome (in his case, revolution) and who saw the purpose of science to be the stripping away of ideological masks to expose the real needs of real men and women. The connection between Peircean pragmatism and Marxist historical materialism, a connection founded on the common ground of praxis, is an important one. Keeping it in mind (even while holding Marxism at a distance) can help us to avoid the cash-value ideology of James's interpreters, especially Rorty—the notion that if something works and endures, then it must be good, a notion that always pleases those who undeservedly hold positions of power over others. This is a reading of James that politically sensitive commentators from John Dewey [17, pp. 103-133] to Frank Lentricchia have debunked [18;19, pp. 62-63]. But it is nevertheless a frequent impression gleaned from the canon of Jamesian pragmatism, an impression of a way of thought that Rorty seems to accept in such statements as this: "we should be more willing than we are to celebrate bourgeois capitalist society as the best polity actualized so far, while regretting that it is irrelevant to most of the problems of most of the population of the planet" [20, p. 210]. And this: "having admitted that [American

society is] a racist, sexist, imperialist society, it seems to me it's still the best thing on offer" [21, p. 9]. A realist version of pragmatism would recognize that the presence of racism, sexism, and imperialism within a seemingly efficient system represents far more than some troublesome noise that has begun to compete with the celebratory music of liberalism; it represents instead real inefficiencies and contradictions that foretell the inability of the system (or political paradigm) to deliver what it has promised or predicted. From the perspective of reformers like Dewey, these social inequalities are problems that need to be solved. Despite the shared interest in praxis and "the real," however, the connection between pragmatism and Marxism cannot be pushed too far. The problem with Marx's analysis of the relation of people to their world lies, as Habermas keenly notes, in the failure to "explicate the interrelationship of interaction and labor," which results in an analysis that "under the unspecific title of social praxis, reduces the one to the other, namely: communicative action to instrumental action" [22, p. 169]. As a corrective to this tendency toward reductiveness in orthodox Marxism, Habermas preserved in his early work an Hegelian orientation to the relation of knowledge and action: "The only knowledge that can truly orient action is knowledge that frees itself from mere human interests and is based on Ideas—in other words, knowledge that has taken a theoretical attitude" [16, p. 301]. Habermas differed from Hegel in his ideas on *how* to achieve the liberation from mere subjective interest, however. His inquiry into this question turned him away from Marx and Hegel and toward the Anglo-American tradition of praxis philosophy. In Peirce's theory of scientific inquiry, Habermas discovered a form of reasoning more to his liking. It is based essentially on the notion that meaning consists of a recognition of the practical outcomes toward which our thinking proceeds. The designation of outcomes, the methods by which we arrive at them, and their overall value are judged not by universal criteria, but by rules established within the community of inquirers in which we work. More than the emphasis on action, it was this emphasis on the normative function of a community of investigators— an emphasis all but ignored by James—that drew Habermas to Peirce's work. The community thus mediates the interaction of the individual mind and the real world and thus sets the pace and purpose of inquiry.

[5]Hacking thus maintains the general social potential we have been seeking in theory. Whereas *thinking* is usually construed as an individual pursuit (and is in fact denigrated as "ideology" or "groupthink" when produced in a communal context), *doing* depends upon interactions with nature and with other people (doing to, doing with).

[6]In his closely related theory of pragmatics, Leech confronts this possibility squarely: "There is likely to be one dissatisfaction with the . . . interpretation of linguistic behavior in terms of means-end analysis. It is that such analysis appears to regard all uses of language as having an instrumental function. Surely, it might be argued, we cannot treat all discourse as motivated by the goal of bringing about some result in the mental or physical condition of the addressee? While one cannot, of course, rule out the occurrence of non-communicative uses of language (e.g., purely expressive speech), it is indeed my contention that, broadly interpreted, the means-end analysis applies to communicative uses of language in general" [36, p. 40].

REFERENCES

1. W. H. Beale, *A Pragmatic Theory of Rhetoric*, Southern Illinois University Press, Carbondale, 1987.

2. J. Kinneavy, *A Theory of Discourse*, Norton, New York, 1971.
3. J. Berlin, *Rhetoric and Reality: Writing Instruction in American Colleges, 1900-1985*, Southern Illinois University Press, Carbondale, 1987.
4. Hayes Microcomputer Products, *User's Manual for Please*, Hayes, Norcross, Georgia, 1984.
5. M. Weber, *Economy and Society: An Outline of Interpretive Sociology*, University of California Press, Berkeley, 1978.
6. H. W. Cummings, L. W. Long, and M. L. Lewis, *Managing Communication in Organizations: An Introduction* (2nd Edition), Gorsuch Scarisbrick, Scottsdale, Arizona, 1987.
7. M. Weber, *The Protestant Ethic and the Spirit of Capitalism*, T. Parsons (trans.), Charles Scribner's Sons, New York, 1958.
8. M. Horkheimer, *Critique of Instrumental Reason*, M. J. O'Connell et al. (trans.), Seabury, New York, 1974.
9. H. Marcuse, *One-Dimensional Man*, Beacon, Boston, 1964.
10. M. Poster, *Critical Theory and Poststructuralism: In Search of a Context*, Cornell University Press, Ithaca, New York, 1989.
11. M. J. Killingsworth and J. S. Palmer, *Ecospeak: Rhetoric and Environmental Politics*, Southern Illinois University Press, Carbondale, 1992.
12. J. Habermas, *The Theory of Communicative Action, Volume 1: Reason and Rationalization of Society*, T. McCarthy (trans.), Beacon, Boston, 1981.
13. P. L. Berger and T. Luckmann, *The Social Construction of Reality: A Treatise in the Sociology of Knowledge*, Doubleday, New York, 1967.
14. J. Habermas, *The Theory of Communicative Action, Volume 2: Lifeworld and System: A Critique of Functionalist Reason*, T. McCarthy (trans.), Beacon, Boston, 1987.
15. I. Illich, *Tools for Conviviality*, Harper and Row, New York, 1973.
16. J. Habermas, *Knowledge and Human Interests*, J. J. Shapiro (trans.), Beacon, Boston, 1971.
17. J. Dewey, *The Quest for Certainty*, G. P. Putnam's Sons, New York, 1929.
18. F. Lentricchia, *Ariel and the Police: Michel Foucault, William James, Wallace Stevens*, University of Wisconsin Press, Madison, 1988.
19. I. Hacking, *Representing and Intervening: Introductory Topics in the Philosophy of Natural Science*, Cambridge University Press, Cambridge, 1983.
20. R. Rorty, *Consequences of Pragmatism*, University of Minnesota Press, Minneapolis, 1982.
21. G. A. Olson, Social Construction and Composition Theory: A Conversation with Richard Rorty, *Journal of Advanced Composition, 9*, pp. 1-9, 1989.
22. J. Habermas, *Theory and Practice*, J. Viertel (trans.), Beacon, Boston, 1973.
23. H. Putnam, *The Many Faces of Realism*, Open Court, La Salle, Illinois, 1987.
24. L. Fleck, *Genesis and Development of a Scientific Fact*, University of Chicago Press, Chicago, 1979.
25. C. Bazerman, *Shaping Written Knowledge: The Genre and Activity of the Experimental Article in Science*, University of Wisconsin Press, Madison, 1988.
26. J. R. Searle, *Speech Acts: An Essay in the Philosophy of Language*, Cambridge University Press, Cambridge, 1969.

27. J. L. Austin, *How to Do Things with Words,* Harvard University Press, Cambridge, Massachusetts, 1962.
28. M. C. McGee, Against Transcendentalism: Prologue to a Functional Theory of Communicative Praxis, in *Politically Speaking: Form and Genre of Political Discourse,* H. W. Simon and A. A. Aghazarian (eds.), University of South Carolina Press, Columbia, 1988.
29. R. W. Dasenbrock, J. L. Austin and the Articulation of a New Rhetoric, *College Composition and Communication, 38,* pp. 291-305, 1987.
30. M. P. Haselkorn, Linguistic Boundaries of Technical Writing, *Technical Writing Teacher, 11,* pp. 26-30, 1983.
31. M. P. Haselkorn, A Pragmatic Approach to Technical Writing, *Technical Writing Teacher, 12,* pp. 122-134, 1984.
32. M. Limaye and R. Cherry, Pragmatics, "Situated Language," and Business Communication, *Journal of Business and Technical Communication, 1,* pp. 68-88, 1987.
33. K. Riley, Conversational Implicature and Unstated Meaning in Professional Communication, *Technical Writing Teacher, 15,* pp. 94-108, 1988.
34. K. Riley, Pragmatics and Technical Communication: Some Further Implications, *Technical Writing Teacher, 13,* pp. 160-170, 1986.
35. K. Riley, Speech Act Theory and Degrees of Directness in Technical Writing, *Technical Writing Teacher, 15,* pp. 1-29, 1988.
36. G. Leech, *Principles of Pragmatics,* Longman, London, 1983.
37. L. A. Hickman, *John Dewey's Pragmatic Technology,* Indiana University Press, Bloomington, 1990.
38. M. Dean, Make Your Reader the Hero of Your How-To Story, *ITCC Proceedings, 30,* pp. WE-3-6, 1983.
39. R. Rorty, *Philosophy and the Mirror of Nature,* Princeton University Press, Princeton, 1979.
40. M. Whitburn, Personality in Scientific and Technical Writing, *Journal of Technical Writing and Communication, 6,* pp. 299-306, 1976.
41. D. Bradford, The Persona in Microcomputer Documentation, *IEEE Transactions on Professional Communication, 27,* pp. 65-68, 1984.
42. B. Rubens, Personality in Computer Documentation: A Preference Study, *IEEE Transactions on Professional Communication, 29,* pp. 56-60, 1986.
43. E. F. Schumacher, *Small is Beautiful: A Study of Economics as if People Mattered,* Blond and Briggs, London, 1973.
44. L. Mumford, *The Myth of the Machine: The Pentagon of Power,* Harcourt, Brace, New York, 1970.
45. L. Mumford, *The Myth of the Machine: Technics and Human Development,* Harcourt, Brace, New York, 1967.
46. W. Whitman, *Complete Poetry and Collected Prose,* Library of America, New York, 1982.
47. M. J. Killingsworth, *Whitman's Poetry of the Body: Sexuality, Politics, and the Text,* University of North Carolina Press, Chapel Hill, 1989.
48. K. Burke, *A Grammar of Motives,* University of California Press, Berkeley, 1945.
49. R. Martin, The New Black Aesthetic Critics and Their Exclusion from American "Mainstream" Criticism, *College English, 50,* pp. 373-382, 1988.

50. R. Martin, *Ishmael Reed and the New Black Aesthetic,* St. Martin's, New York, 1988.
51. M. L. King, Jr., Letter from the Birmingham Jail, in *The Norton Reader* (5th Edition), A. M. Eastman, et al. (eds.), Norton, New York, pp. 852-866, 1980.

Bibliography

Allen, J., Breaking with a Tradition: New Directions in Audience Analysis, *Technical Writing: Theory and Practice,* B. E. Fearing and W. K. Sparrow (eds.), Modern Language Association, New York, pp. 53-71, 1989.

Anderson, P. V., *Technical Writing: A Reader-Centered Approach,* Harcourt Brace Javonovitch, New York, 1987.

Anderson, S. A., Managing through Automation, *ITCC Proceedings, 33,* pp. 118-120, 1986.

Anon., *Diamond Sutra,* A. F. Price and W. Mou-Lam (trans.), Shambhala, Boulder, Colorado, 1969.

Anon., *Nota Bene Tutorial for Version 2.0,* Equal Access Systems, 1986.

Aristotle, *Rhetoric,* in *The Complete Works of Aristotle: The Revised Oxford Translation,* 2 vol., J. Barnes (ed.), Princeton University Press, Princeton, 1984.

Arnheim, R., *Visual Thinking,* University of California Press, Berkeley, 1969.

Austin, J. L., *How to Do Things with Words,* Harvard University Press, Cambridge, Massachusetts, 1962.

Barker, T. T. (ed.), *Perspectives on Software Documentation: Inquiries and Innovations,* Baywood Publishing Company, Amityville, New York, 1990.

Barstow, T. R., and J. T. Jaynes, Integrating Online Documentation into the Technical Publishing Process, *IEEE Transactions on Professional Communication, 29,* pp. 37-41, 1986.

Barthes, R., *Elements of Semiology,* A. Lavers and C. Smith (trans.), Hill and Wang, New York, 1968.

Barthes, R., *Image, Music, Text,* S. Heath (trans.), Hill and Wang, New York, 1977.

Barthes, R., *Mythologies,* A. Lavers (trans.), Hill and Wang, New York, 1972.

Bartholomae, D., Inventing the University, in *When a Writer Can't Write,* M. Rose (ed.), Guilford, New York, pp. 134-165, 1985.

Barton, B. F., and M. S. Barton, Toward a Rhetoric of Visuals for the Computer Age, *Technical Writing Teacher, 12,* pp. 126-145, 1985.

Barzun, J., and H. F. Graff, *The Modern Researcher*, Harcourt Brace Javonovitch, New York, 1970.

Bateson, G., *Steps to an Ecology of Mind*, Granada, London, 1973.

Bazerman, C., *Shaping Written Knowledge: The Genre and Activity of the Experimental Article in Science*, University of Wisconsin Press, Madison, 1988.

Beale, W. H., *A Pragmatic Theory of Rhetoric*, Southern Illinois University Press, Carbondale, 1987.

Beatts, P. M., Use Your Reader's Eyes, *IRE Transactions on Engineering Writing and Speech, 2*, pp. 6-11, 1959.

Beene, L., and P. White (eds.), *Solving Problems in Technical Writing*, Oxford University Press, New York, 1988.

Benveniste, E., *Problems in General Linguistics*, M. E. Meek (trans.), University of Miami Press, Coral Gables, Florida, 1971.

Berger, P. L., and T. Luckmann, *The Social Construction of Reality: A Treatise in the Sociology of Knowledge*, Doubleday, New York, 1967.

Berlin, J., *Rhetoric and Reality: Writing Instruction in American Colleges, 1890-1985*, Southern Illinois University Press, Carbondale, 1987.

Bernstein, R., *Praxis and Action: Contemporary Philosophies of Human Action*, University of Pennsylvania Press, Philadelphia, 1971.

Bertin, J., *Graphics and Graphic Information Processing*, W. Berg and P. Scott (trans.), Walter de Gruyter, New York, 1981.

Bizzell, P., Cognition, Convention, and Certainty, *Pre/Text, 3*, pp. 213-243, 1982.

Bizzell, P., Foundationalism and Anti-Foundationalism in Composition Studies, *Pre/Text, 7*, pp. 27-56, 1986.

Boorstin, D. J., *The Discoverers*, Vintage Books, New York, 1983.

Bradford, D., The Persona in Microcomputer Documentation, *IEEE Transactions on Professional Communication, 27*, pp. 65-68, 1984.

Broadhead, G. J., and R. C. Freed, *The Variables of Composition: Process and Product in a Business Setting*, Southern Illinois University Press, Carbondale, 1986.

Brufee, K. A., Social Construction, Language, and the Authority of Knowledge: A Bibliographical Essay, *College English, 48*, pp. 773-790, 1986.

Burke, K., *Counter-Statement*, Harcourt, New York, 1931.

Burke, K., *Grammar of Motives*, University of California Press, Berkeley, 1945.

Burke, K., *A Rhetoric of Motives*, University of California Press, Berkeley, 1950.

Campbell, C. R., What Can Discourse Theory Offer the Professional Who Writes?, *IEEE Transactions in Professional Communication, 33*, pp. 156-161, 1990.

Capra, F., *The Turning Point: Science, Society, and the Rising Culture*, Simon and Schuster, New York, 1982.

Chavarria, L. S., Improving the Friendliness of Technical Manuals, *ITCC Proceedings, 29*, pp. W26-28, 1982.

[Chew, J.,] *Interbridge System Guide*, Hayes, Norcross, Georgia, 1986.

Chew, J., J. Jandel, and A. Martinich, The New Communicator: Engineering the Human Factor, *ITCC Proceedings, 33*, pp. 422-423, 1986.

Cole, R., Work Reform and Quality Circles in Japanese Industry, in *Critical Studies in Organization and Bureaucracy*, F. Fischer and C. Sirianni (eds.), Temple University Press, Philadelphia, pp. 421-452, 1984.

Cooley, M., *Architect or Bee?: The Human/Technology Relationship*, South End Press, Boston, 1980.

Cooper, M., The Ecology of Writing, *College English, 48*, pp. 364-375, 1986.

Connors, R. J., The Rise of Technical Writing Instruction in America, *Journal of Technical Writing and Communication, 12*, pp. 333-337, 1982.

Corbett, E. P. G., *Classical Rhetoric for the Modern Student* (2nd Edition), Oxford University Press, New York, 1971.

Cummings, H. W., L. W. Long, and M. L. Lewis, *Managing Communication in Organizations: An Introduction* (2nd Edition), Gorsuch Scarisbrick, Scottsdale, Arizona, 1987.

Cunningham, D. H., and G. Cohen, *Creating Technical Manuals*, McGraw-Hill, New York, 1984.

D'Angelo, F., *Process and Thought in Composition* (3rd Edition), Little, Brown, Boston, 1985.

Dasenbrock, R. W., J. L. Austin and the Articulation of a New Rhetoric, *College Composition and Communication, 38*, pp. 291-305, 1987.

Dean, M., Make Your Reader the Hero of Your How-To Story, *ITCC Proceedings, 30*, pp. WE-3-6, 1983.

Debs, M. B., Collaborative Writing in Industry, in *Technical Writing: Theory and Practice*, B. Fearing and K. Sparrow (eds.), Modern Language Association, New York, pp. 33-42, 1989.

Derrida, J., *Of Grammatology*, G. Spivak (trans.), Johns Hopkins University Press, Baltimore, 1976.

Dewey, J., *The Quest for Certainty*, G. P. Putnam's Sons, New York, 1929.

Diamond, J., Soft Sciences Are Often Harder than Hard Sciences, pp. 34-39, August, 1987.

Dilbeck, A. M., and l. A. Golowich, Developing the Integrated Documentation Team, *ITCC Proceedings, 35*, pp. MPD 31-33, 1988.

Dobrin, D. N., *Writing and Technique*, National Council of Teachers of English, Urbana, Illinois, 1989.

Doheny-Farina, S., and L. Odell, Ethnographic Research on Writing: Assumptions and Methodology, in *Writing in Nonacademic Settings*, L. Odell and D. Goswami (eds.), Guilford, New York, pp. 503-535, 1985.

Dragga, S., and G. Gong, *Editing: The Design of Rhetoric*, Baywood, Amityville, New York, 1989.

Dressel, S., J. S. Euler, S. A Bagby, and S. A. Dell, ASAPP: Automated Systems Approach to Proposal Production, *IEEE Transactions on Professional Communication, 26,* pp. 63-68, 1983.

Eco, U., *A Theory of Semiotics,* Indiana University Press, Bloomington, 1979.

Elliott, R. C., *The Literary Persona,* University of Chicago Press, Chicago, 1982.

Emig, J., *The Composing Processes of Twelfth Graders,* NCTE, Urbana, Illinois, 1971.

Englebert, D., Storyboarding—A Better Way of Planning and Writing Proposals, *IEEE Transactions on Professional Communication, 15,* pp. 115-118, 1972.

Enrick, N. L., *Effective Graphic Communication,* Auerbach, New York, 1972.

Evans, F. J., Jr., More Effective Engineering Proposals, *IRE Transactions on Engineering Writing and Speech, 3,* pp. 54-58, 1960.

Faigley, L., Competing Theories of Process: A Critique and a Proposal, *College English, 48,* pp. 773-790, 1986.

Faigley, L., Nonacademic Writing: The Social Perspective, in *Writing in Nonacademic Settings,* L. Odell and D. Goswami (eds.), Guilford, New York, pp. 231-248, 1985.

Federal Express Corporation, Annual Report, Memphis, Tennessee, 1988.

Fisch, M., *Peirce, Semeiotic, and Pragmatism: Essays by Max Fisch,* K. L. Ketner and C. J. W. Kloesel (eds.), Indiana University Press, Bloomington, 1986.

Fish, S., *Is There a Text in This Class? The Authority of Interpretive Communities,* Harvard University Press, Cambridge, Massachusetts, 1980.

Fleck, L., *Genesis and Development of a Scientific Fact,* University of Chicago Press, Chicago, 1979.

Flower, L., and J. R. Hayes, A Cognitive Process Theory of Writing, *College Composition and Communication, 32,* pp. 365-387, 1981.

Flower, L., and J. R. Hayes, Problem-Solving Strategies and the Writing Process, *College English, 39,* pp. 449-461, 1977.

Flower, L., J. Hayes, and H. Swarts, Revising Functional Documents: The Scenario Principle, in *New Essays in Technical and Scientific Communication: Research, Theory, Practice,* P. V. Anderson, R. J. Brockmann, and C. M. Miller (eds.), Baywood, Amityville, New York, pp. 41-58, 1983.

Fowler, S. L., and D. Roeger, Programmer and Writer Collaboration: Making User Manuals that Work, *IEEE Transactions on Professional Communication, 29,* pp. 21-25, 1986.

Freed, R. C., and G. J. Broadhead, Discourse Communities, Sacred Texts, and Institutional Norms, *College Composition and Communication, 38,* pp. 154-165, 1987.

Frye, N., *Anatomy of Criticism,* Princeton University Press, Princeton, 1957.

Geertz, C., *The Interpretation of Cultures,* Basic Books, New York, 1973.

Geertz, C., *Local Knowledge: Further Essays in Interpretive Anthropology,* Basic Books, New York, 1983.

Gilbertson, M. K., and M. J. Killingsworth, Classical Rhetoric and Technical Communication, in *Teaching the History and Rhetoric of Scientific and Technical Literature*, S. Southard (ed.), Association of Teachers of Technical Writing Anthology Series, ATTW, forthcoming.

Gould, J. M., *The Technical Elite*, Augustus M. Kelley, New York, 1966.

Green, R., The Graphic-Oriented (GO) Proposal Primer, *ITCC Proceedings, 32*, pp. VC 29-30, 1985.

Grice, R. A., Document Development in Industry, in *Technical Writing: Theory and Practice*, B. Fearing and K. Sparrow (eds.), Modern Language Association, New York, pp. 27-32, 1989.

Grimm, S. J., *How to Write Computer Manuals for Users*, Lifetime Learning Publications, Belmont, California, 1982.

Gross, A., *The Rhetoric of Science*, Harvard University Press, Cambridge, Massachusetts, 1990.

Gumperz, J., The Speech Community, in *Language and Social Context*, P. P. Giglioli (ed.), Penguin, Baltimore, pp. 219-231, 1972.

Habermas, J., *Knowledge and Human Interests*, J. Shapiro (trans.), Beacon, Boston, 1971.

Habermas, J., *Theory and Practice*, J. Viertel (trans.), Beacon, Boston, 1973.

Habermas, J., *The Theory of Communicative Action, Volume 1: Reason and Rationalization of Society*, T. McCarthy (trans.), Beacon, Boston, 1981.

Habermas, J., *The Theory of Communicative Action, Volume 2: Lifeworld and System: A Critique of Functionalist Reason*, T. McCarthy (trans.), Beacon, Boston, 1987.

Hacking, I., *Representing and Intervening: Introductory Topics in the Philosophy of Natural Science*, Cambridge University Press, Cambridge, 1983.

Hackman, J. R., The Design of Self-Managing Work Groups, in *Managerial Control and Organizational Democracy*, B. King, S. Streufert, and F. E. Fredler (eds.), Wiley, New York, pp. 61-89, 1978.

Hackman, J. R., and G. R. Oldham, Motivation through the Design of Work: Test of a Theory, *Organizational Behavior and Human Performance, 16*, pp. 260-279, 1976.

Hairston, M., Winds of Change: Thomas Kuhn and the Revolution in the Teaching of Writing, *College Composition and Communication, 33*, pp. 76-88, 1982.

Halloran, S. M., Technical Writing and the Rhetoric of Science, *Journal of Technical Writing and Communication, 8*, pp. 77-88, 1979.

Halloran, S. M., and M. D. Whitburn, Ciceronian Rhetoric and the Rise of Science: The Plain Style Reconsidered, in *The Rhetorical Tradition and Modern Writing*, J. J. Murphy (ed.), Modern Language Association, New York, pp. 58-72, 1982.

Harris, J., The Idea of Community in the Study of Writing, *College Composition and Communication, 40*, pp. 11-22, 1989.

Hart, R. P., *Modern Rhetorical Criticism,* Scott, Foresman/Little, Brown, Glenview, Illinois, 1990.

Haselkorn, M. P., Linguistic Boundaries of Technical Writing, *Technical Writing Teacher, 11,* pp. 26-30, 1983.

Haselkorn, M. P., A Pragmatic Approach to Technical Writing, *Technical Writing Teacher, 11,* pp. 122-134, 1984.

Hayes Microcomputer Products, User's Manual for *Please,* Hayes, Norcross, Georgia, 1984.

Hickman, L. A., *John Dewey's Pragmatic Technology,* Indiana University Press, Bloomington, 1990.

Hill, F. I., The *Rhetoric* of Aristotle, in *A Synoptic History of Classical Rhetoric,* J. J. Murphy (ed.), Hermagoras Press, Davis, California, pp. 19-76, 1983.

Hodge, R., and G. Kress, *Social Semiotics,* Cornell University Press, Ithaca, New York, 1988.

Holtz, H., *Government Contracts: Proposalmanship and Winning Strategies,* Plenum, New York, 1979.

Holtz, H., and T. Schmidt, *The Winning Proposal—How to Write It,* McGraw-Hill, New York, 1981.

Horkheimer, M., *Critique of Instrumental Reason,* M. J. O'Connell et al. (trans.), Seabury, New York, 1974.

Houp, K. W., and T. E. Pearsall, *Reporting Technical Information,* Macmillan, New York, 1968.

Hunter, L., A Rhetoric of Mass Communication: Collective or Corporate Public Discourse, in *Oral and Written Communication: Historical Approaches,* R. L. Enos (eds.), Sage Publications, Newbury Park, California, pp. 216-261, 1990.

Ihde, D., *Technology and the Lifeworld: From Garden to World,* Indiana University Press, Bloomington, 1990.

Illich, I., *Tools for Conviviality,* Harper and Row, New York, 1973.

Jakobson, R., Shifters, Verbal Categories, and the Russian Verb, in *Selected Writings of Roman Jakobson,* Mouton, The Hague, Vol. 2, pp. 130-147, 1971.

Jesperson, O., *Language: Its Nature, Development and Origin,* Norton, New York, 1964.

Kallendorf, C., and C. Kallendorf, Aristotle and the Ethics of Business Communication, *Iowa State Journal of Business and Technical Communication, 3,* pp. 54-69, 1989.

Kalmach, J., Technical Writing Teachers and the Challenges of Desktop Publishing, *The Technical Writing Teacher, 15,* pp. 119-131, 1988.

Kanter, R. M., Empowerment, in *The Leader-Manager,* J. N. Williamson (ed.), Wiley, New York, pp. 479-504, 1986.

Keene, M., and M. Barnes-Ostrander, Audience Analysis and Adaptation, in *Research in Technical Communication: A Bibliographic Sourcebook,*

M. Moran and D. Journet (eds.), Greenwood, Westport, Connecticut, pp. 163-191, 1985.

Kenney, M., *Biotechnology: The University-Industrial Complex,* Yale University Press, New Haven, 1986.

Killingsworth, M. J., The Essay and the Report: Expository Poles in Technical Writing, *Journal of Technical Writing and Communication, 15,* pp. 227-233, 1985.

Killingsworth, M. J., How to Talk about Professional Communication: Metalanguage and Heuristic Power, *Journal of Business and Technical Communication, 3,* pp. 117-125, 1989.

Killingsworth, M. J., Science and Technology for a General Audience: The Personal Essay, *IEEE Transactions in Professional Communication, 25,* pp. 186-188, 1982.

Killingsworth, M. J., Toward a Rhetoric of Technological Action, *ITCC Proceedings, 34,* pp. RET-136-39, 1987.

Killingsworth, M. J., *Whitman's Poetry of the Body: Sexuality, Politics, and the Text,* University of North Carolina Press, Chapel Hill, 1989.

Killingsworth, M. J., and K. Eiland, Managing the Production of Technical Manuals: Recent Trends, *IEEE Transactions on Professional Communication, 29,* pp. 23-26, 1986.

Killingsworth, M. J., and M. K. Gilbertson, Rhetoric and Relevance in Technical Writing, *Journal of Technical Writing and Communication, 16,* pp. 287-297, 1986.

Killingsworth, M. J., M. K. Gilbertson, and J. Chew, Amplification in Technical Manuals: Theory and Practice, *Journal of Technical Writing and Communication, 19,* pp. 13-29, 1989.

Killingsworth, M. J., and B. G. Jones, Division of Labor or Integrated Teams: A Crux in the Management of Technical Communication?, *Technical Communication, 36,* pp. 210-221, 1989.

Killingsworth, M. J., and J. S. Palmer, *Ecospeak: Rhetoric and Environmental Politics in America,* Southern Illinois University Press, Carbondale, 1992.

Killingsworth, M. J., and D. Steffens, Effectiveness in the Environmental Impact Statement: A Study in Public Rhetoric, *Written Communication, 6,* pp. 155-180, 1989.

King, M. L., Jr., Letter from the Birmingham Jail, in *The Norton Reader* (5th Edition), Arthur M. Eastman et al. (eds.), Norton, New York, pp. 852-866, 1980.

Kinneavy, J., *A Theory of Discourse,* Norton, New York, 1971.

Knuth, D. E., *The TeXbook,* Addison Wesley, Reading, Massachusetts, 1986.

Krull, R., and J. M. Hurford, Can Computers Increase Writing Productivity?, *Technical Communication, 34,* pp. 243-249, 1987.

Kuhn, T. S., *The Structure of Scientific Revolutions* (2nd Edition), University of Chicago Press, Chicago, 1970.

Langer, S., *Philosophy in a New Key,* Harvard University Press, Cambridge, Massachusetts, 1942.

Lanham, R., *Analyzing Prose,* Charles Scribner's Sons, New York, 1983.

Lauer, J. M., and J. W. Asher, *Composition Research: Empirical Designs,* Oxford University Press, New York, 1988.

Leech, G., *Principles of Pragmatics,* Longman, London, 1983.

LeFevre, K. B., *Invention as a Social Process,* Southern Illinois University Press, Carbondale, 1987.

Lefferts, R., *Elements of Graphics: How to Prepare Charts and Graphs for Effective Reports,* Harper and Row, New York, 1981.

Lentricchia, F., *Ariel and the Police: Michel Foucault, William James, Wallace Stevens,* University of Wisconsin Press, Madison, 1988.

Limaye, M., and R. Cherry, Pragmatics, "Situated Language," and Business Communication, *Journal of Business and Technical Communication, 1,* pp. 68-88, 1987.

Long, R., Writer-Audience Relationships: Analysis or Invention?, *College Composition and Communication, 31,* pp. 221-226, 1980.

Lyne, J. R., Rhetoric and Semiotic, *Quarterly Journal of Speech, 66,* pp. 155-168, 1980.

Machlup, F., *The Production and Distribution of Knowledge in the United States,* Princeton University Press, Princeton, New Jersey, 1962.

MacIntire, K., "Leonardo Da Vinci: Technical Communicator," M.A. Thesis, Memphis State University, 1989.

Magnan, G., Industrial Illustrating and Layout, in *Handbook of Technical Writing Practices,* S. Jordon, J. Kleinman, and H. Shimberg (eds.), Wiley and Sons, vol. 2, pp. 735-800, 1971.

Manyak, T. G., The Management of Business Writing, *Journal of Technical Writing and Communication, 16,* pp. 355-361, 1986.

Marcuse, H., *One-Dimensional Man,* Beacon, Boston, 1964.

Martin, R., *Ishmael Reed and the New Black Aesthetic,* St. Martin's, New York, 1988.

Martin, R., The New Black Aesthetic Critics and Their Exclusion from American "Mainstream" Criticism, *College English, 50,* pp. 373-382, 1988.

Marx, K., and F. Engels, *The German Ideology,* R. Pascal (ed.), International Publishers, New York, 1947.

Masse, R. E., and M. D. Benz, Technical Communication and Rhetoric, in *Technical and Business Communication: Bibliographic Essays for Teachers and Corporate Trainers,* C. H. Sides (ed.), National Council of Teachers of English, Urbana, Illinois, pp. 5-38, 1989.

Matalene, C. B. (ed.), *Worlds of Writing: Teaching and Learning in Discourse Communities at Work,* Random House, New York, 1989.

Maynard, J., A User Driven Approach to Better User Manuals, *IEEE Transactions on Professional Communication, 25,* pp. 16-19, 1982.

Mayo, E., *The Human Problems of an Industrial Civilization*, Macmillan, New York, 1933.

McGee, M. C., Against Transcendentalism: Prologue to a Functional Theory of Communicative Praxis, in *Politically Speaking: Form and Genre of Political Discourse*, H. W. Simon and A. A. Aghazarian (eds.), University of South Carolina Press, Columbia, 1988.

McLuhan, M., *Understanding Media: The Extensions of Man*, McGraw-Hill, New York, 1964.

Mead, G. H., *Mind, Self, and Society*, C. W. Morris (ed.), University of Chicago Press, Chicago, 1934.

Mertz, E., Beyond Symbolic Anthropology: Introducing Semiotic Mediation, in *Semiotic Mediation: Sociological and Psychological Perspectives*, E. Mertz and R. J. Parmentier (eds.), Academic Press, Orlando, pp. 1-19, 1985.

Miller, C. R., Genre as Social Action, *Quarterly Journal of Speech, 70*, pp. 151-167, 1984.

Miller, C. R., A Humanistic Rationale for Technical Writing, *College English, 40*, pp. 610-617, 1979.

Milner, W., The Technical Writer's Role in On-Line Documentation, *ITCC Proceedings, 33*, pp. 61-64, 1986.

Mumford, L., *The Myth of the Machine: The Pentagon of Power*, Harcourt, Brace, New York, 1970.

Mumford, L., *The Myth of the Machine: Technics and Human Development*, Harcourt, Brace, New York, 1967.

Murdich, R. J., How to Evaluate Engineering Proposals, *Machine Design, 33*, pp. 116-120, 1961.

Noble, D. F., *America by Design: Science, Technology, and the Rise of Corporate Capitalism*, Oxford University Press, New York, 1977.

Ogilvy, D., *Confessions of an Advertising Man*, Atheneum, New York, 1964.

Ogilvy, D., *Ogilvy on Advertising*, Crown, New York, 1983.

Ohmann, R., *English in America: A Radical View of the Profession*, Oxford University Press, New York, 1976.

Olson, G. A., Social Construction and Composition Theory: A Conversation with Richard Rorty, *Journal of Advanced Composition, 9*, pp. 1-9, 1989.

Ong, W. J., Voice as a Summons for Belief, *The Barbarian Within*, Macmillan, New York, pp. 49-67, 1962.

Oslund, R. R., Brochuremanship—How Much Is Too Much?, *STWP Review, 12*:3, pp. 14-15, 1965.

Oslund, R. R., Brochuremanship Versus Cost, *Data, 7*, pp. 59-60, 1962.

Packard, V., *The Hidden Persuaders*, D. McKay, New York, 1957.

Park, D. B., Analyzing Audiences, *College Composition and Communication, 37*, pp. 478-488, 1986.

Parker, A., Problem Solving Applied to Teaching Technical Writing, *The Technical Writing Teacher, 17*, pp. 95-103, 1990.

Pasmore, W., C. Francis, J. Haldeman, and A. Shani, Sociotechnical Systems: A North American Reflection on Empirical Studies of the Seventies, *Human Relations, 35,* pp. 1179-1204, 1982.

Pearsall, T. E., *Audience Analysis for Technical Writing,* Glencoe, Beverly Hills, California, 1969.

Pearsall, T. E., Panel Presentation on Technical Communication, Rocky Mountain Modern Language Association, El Paso, 1984.

Pearsall, T. E., and D. Cunningham, *How to Write for the World of Work* (3rd Edition), Holt, Rinehart and Winston, New York, 1984.

Peirce, C. S., *Classical American Philosophy: Essential Readings and Interpretive Essays,* J. J. Stuhr (ed.), Oxford University Press, New York, 1987.

Peirce, C. S., *The Collected Papers of Charles Sanders Peirce,* 5 vols., C. Hartshorne and P. Weiss (eds.), Harvard University Press, Cambridge, Massachusetts, 1960.

Peirce, C. S., *Semiotic and Significs: The Correspondence of Charles S. Peirce and Lady Victoria Welby,* C. S. Hardwick (ed.), Indiana University Press, Bloomington, 1977.

Pike, K., *Linguistic Concepts: An Introduction to Tagmemics,* University of Nebraska Press, Lincoln, 1982.

Pinelli, T. E., T. E. Pearsall, and R. A. Grice, Introduction to Special Issue on Productivity Management and Enhancement in Technical Communication, *Technical Communication, 34,* pp. 216-218, 1987.

Popper, K. R., *The Open Society and Its Enemies,* 2 vols., Princeton University Press, Princeton, New Jersey, 1966.

Porter, J. E., Intertextuality and the Discourse Community, *Rhetoric Review, 5,* pp. 34-47, 1986.

Poster, M., *Critical Theory and Poststructuralism: In Search of a Context,* Cornell University Press, Ithaca, New York, 1989.

Poza, E. J., and M. L. Markus, Success Story: The Team Approach to Work Restructuring, *Organizational Dynamics, 25,* pp. 3-25, 1980.

Price, D. J. de Solla, *Little Science, Big Science . . . and Beyond,* Columbia University Press, New York, 1986.

Price, J., *How to Write a Computer Manual,* Benjamin/Cummings, Menlo Park, California, 1984.

Proietti, H. L., and J. L. Thomas, Multifunctional Team Dynamics: A Success Story, *ITCC Proceedings, 35,* pp. MPD 58-60, 1988.

Putnam, H., *The Many Faces of Realism,* Open Court, La Salle, Illinois, 1987.

Ransdell, J., Peircean Semiotic: The Grammar of Representation, unpublished manuscript, 1988.

Ransdell, J., Some Leading Ideas of Peirce's Semiotic, *Semiotica, 19,* pp. 159-178, 1977.

Redish, J. C., and D. A. Schell, Writing and Testing Instructions for Usability, Technical Writing: Theory and Practice, B. E. Fearing and W. K. Sparrow (eds.), Modern Language Association, New York, pp. 63-71, 1989.

Ricoeur, P., *The Philosophy of Paul Ricoeur: An Anthology of His Work,* Beacon, Boston, 1978.

Riley, K., Conversational Implicature and Unstated Meaning in Professional Communication, *Technical Writing Teacher, 15,* pp. 94-108, 1988.

Riley, K., Pragmatics and Technical Communication: Some Further Implications, *Technical Writing Teacher, 13,* pp. 160-170, 1986.

Riley, K., Speech Act Theory and Degrees of Directness in Technical Writing, *Technical Writing Teacher, 15,* pp. 1-29, 1988.

Rochberg-Halton, E., *Meaning and Modernity: Social Theory in the Pragmatic Attitude,* University of Chicago Press, Chicago, 1986.

Rorty, R., *Consequences of Pragmatism,* University of Minnesota Press, Minneapolis, 1982.

Rorty, R., *Philosophy and the Mirror of Nature,* Princeton University Press, Princeton, 1979.

Rubens, B., Personality in Computer Documentation: A Preference Study, *IEEE Transactions on Professional Communication, 29,* pp. 56-60, 1986.

Rubens, P. M., Desktop Publishing: Technology and the Technical Communicator, *Technical Communication, 35,* pp. 296-299, 1988.

Rude, C., Format and Typography in Complex Instructions, *ITCC Proceedings, 32,* pp. RET 36-39, 1985.

Russell, C. A., *Science and Social Change in Britain and Europe, 1700-1900,* St. Martin's, New York, 1983.

Sanders, S. P., How Can Technical Writing Be Persuasive?, *Solving Problems in Technical Writing,* L. Beene and P. White (eds.), Oxford University Press, New York, pp. 55-78, 1988.

Scholes, R., *Semiotics and Interpretation,* Yale University Press, New Haven, 1982.

Schumacher, E. F., *Small Is Beautiful: A Study of Economics as if People Mattered,* Blond and Briggs, London, 1973.

Scofield, R. J., Can Graphics Do the Entire Communication Job?, *ITCC Proceedings, 24,* pp. 174-176, 1977.

Searle, J. R., *Speech Acts: An Essay in the Philosophy of Language,* Cambridge University Press, Cambridge, 1969.

Selzer, J., Composing Processes for Technical Discourse, in *Technical Writing: Theory and Practice,* B. Fearing and K. Sparrow (eds.), Modern Language Association, New York, pp. 43-50, 1989.

Siegel, A., The Plain English Revolution, in *Readings in Technical Writing,* D. C. Leonard and P. J. McGuire (eds.), Macmillan, New York, pp. 222-229, 1983.

Silverman, K., *The Subject of Semiotics,* Oxford University Press, New York, 1983.

Skees, W. D., *Writing Handbook for Computer Professionals,* Lifetime Learning Publications, Belmont, California, 1982.

Smallwood, M. S., Including Writers on the Software Development Team, in *ITCC Proceedings, 33,* pp. 387-390, 1986.

Stacy, R. H., *Defamiliarization in Language and Literature,* Syracuse University Press, Syracuse, New York, 1977.

Stoddard, E. W., The Role of Ethos in the Theory of Technical Writing, *The Technical Writing Teacher, 11,* pp. 229-241, 1984.

Stratton, C., *Technical Writing: Process and Product,* Holt, Rinehart, and Winston, New York, 1984.

Sutton, F. X., S. E. Harris, C. Kaysen, and J. Tobin, *The Business Creed,* Harvard University Press, Cambridge, Massachusetts, 1956.

Sykes, C. J., *ProfScam: Professors and the Demise of Higher Education,* St. Martin's, New York, 1988.

Tebeaux, E., Visual Language: The Development of Format and Page Design in the English Renaissance Technical Writing, *Journal of Business and Technical Writing, 5,* pp. 246-274, 1991.

Titen, J., Application of Rudolf Arnheim's Visual Thinking to the Teaching of Technical Writing, *Technical Writing Teacher, 7,* pp. 113-118, 1980.

Tracey, J. J., The Theory and Lessons of STOP Discourse, *IEEE Transactions on Professional Communication, 26,* pp. 68-78, 1983.

Turnbull, A. T., and R. Baird, *The Graphics of Communication* (3rd Edition), Holt, Rinehart, and Winston, New York, 1975.

Volosinov, V. N., *Marxism and the Philosophy of Language,* L. Matejka and I. R. Titunik (trans.), Seminar Press, New York, 1973.

Waller, P. L., "The Role of Ethos in the Writing of Proposals and Manuals," Doctoral Dissertation, Texas Tech University, 1988.

Warnock, J., The Writing Process, in *Research in Composition and Rhetoric: A Bibliographic Sourcebook,* M. Moran and R. Lunsford (eds.), Greenwood Press, Westport, Connecticut, pp. 3-26, 1984.

Warren, T. L., The Passive Voice Verb: An Annotated Bibliography, Parts I and II, *Journal of Technical Writing and Communication, 11,* pp. 271-286, 373-389, 1981.

Weber, M., *Economy and Society: An Outline of Interpretive Sociology,* University of California Press, Berkeley, 1978.

Weber, M., *The Protestant Ethic and the Spirit of Capitalism,* T. Parsons (trans.), Charles Scribner's Sons, New York, 1958.

Weigley, R., *The American Way of War: A History of United States Military Strategy and Policy,* Indiana University Press, Bloomington, 1973.

Whitburn, M., Personality in Scientific and Technical Writing, *Journal of Technical Writing and Communication, 6,* pp. 299-306, 1976.

Whitburn, M., et al., The Plain Style in Technical Writing, *Journal of Technical Writing and Communication, 8,* pp. 349-358, 1978.

White, J. V., *Graphic Design for the Electronic Age: The Manual for Traditional and Desktop Publishing,* Watson-Guptill, New York, 1988.

Whitman, W., *Complete Poetry and Collected Prose,* Library of America, New York, 1982.

Williams, J., *Style, Ten Lessons in Clarity and Grace* (2nd Edition), Scott, Foresman, Glenview, Illinois, 1985.

Wimberly, L. W., The Technical Writer as a Member of the Software Development Team: A Model for Contributing More Professionally, *ITCC Proceedings, 30,* pp. GEP 36-38, 1983.

Winkler, V. M., The Role of Models in Technical and Scientific Writing, in *New Essays in Technical and Scientific Communication: Research, Theory, Practice,* P. V. Anderson, R. J. Brockmann, and C. R. Miller (eds.), Baywood, Amityville, New York, pp. 111-122, 1983.

Young, R., Paradigms and Problems: Needed Research in Rhetorical Invention, in *Research on Composing: Points of Departure,* C. Cooper and L. Odell (eds.), NCTE, Urbana, Illinois, pp. 29-47, 1978.

Zappen, J. P., Rhetoric and Technical Communication: An Argument for Historical and Political Pluralism, *Iowa State Journal of Business and Technical Communication, 1,* pp. 29-44, 1987.

Zoellner, R., Talk-Write: A Behavioral Pedagogy for Composition, *College English, 30,* pp. 267-320, 1969.

Zuboff, S., *In the Age of the Smart Machine: The Future of Work and Power,* Basic Books, New York, 1987.

Index